SpringerBriefs in Applied Sciences and Technology

PoliMI SpringerBriefs

Springer, in cooperation with Politecnico di Milano, publishes the PoliMI SpringerBriefs, concise summaries of cutting-edge research and practical applications across a wide spectrum of fields. Featuring compact volumes of 50 to 125 (150 as a maximum) pages, the series covers a range of contents from professional to academic in the following research areas carried out at Politecnico:

- Aerospace Engineering
- Bioengineering
- Electrical Engineering
- Energy and Nuclear Science and Technology
- Environmental and Infrastructure Engineering
- Industrial Chemistry and Chemical Engineering
- Information Technology
- Management, Economics and Industrial Engineering
- Materials Engineering
- Mathematical Models and Methods in Engineering
- Mechanical Engineering
- Structural Seismic and Geotechnical Engineering
- Built Environment and Construction Engineering
- Physics
- Design and Technologies
- Urban Planning, Design, and Policy

http://www.polimi.it

Mariacristina Giambruno · Sonia Pistidda ·
Nune Petrosyan

Heritage-Based Tools for the Sustainable Urban Development of Historic Centers

The Case Study of Gyumri in Armenia

POLITECNICO
MILANO 1863

Springer

Mariacristina Giambruno
DASTU
Politecnico di Milano
Milan, Italy

Sonia Pistidda
DASTU
Politecnico di Milano
Milan, Italy

Nune Petrosyan
NUACA
Yerevan, Armenia

ISSN 2191-530X ISSN 2191-5318 (electronic)
SpringerBriefs in Applied Sciences and Technology
ISSN 2282-2577 ISSN 2282-2585 (electronic)
PoliMI SpringerBriefs
ISBN 978-3-031-75258-2 ISBN 978-3-031-75259-9 (eBook)
https://doi.org/10.1007/978-3-031-75259-9

This Springer imprint is published by the registered company Springer Nature Switzerland AG
The registered company address is: Gewerbestrasse 11, 6330 Cham, Switzerland

Preface

The rapid growth of cities, driven by globalization and climate change, has led to profound transformations in their cultural assets. While the need to protect monuments is widely recognized, a different sensitivity is often applied to the widespread heritage that comprises historic centers: common houses often built with simple materials and construction techniques, yet crucial to the urban identity of cities. The systemic nature of this heritage helps define the urban backbone and, along with the quality of open spaces, can foster resilient communities. However, urban planning tools are often ill-equipped to manage the complexity of historic cities, prioritizing conservative transformations that respect their layers and uniqueness.

At a time of significant global challenges, the built heritage of these places can play a crucial role in sustainable transformation by optimizing existing resources, reducing waste and energy consumption, and preserving their distinctiveness.

This volume outlines potential steps for conducting preliminary studies and defining urban planning strategies using the 2021 case study of the Armenian city of Gyumri. The approach emphasizes not only the conservation of built heritage but also the creation of a vibrant and dynamic core for the contemporary city.

Gyumri and its historic center (the Kumayri ancient core) are unique cases in the Armenian context in terms of size, building characteristics, and urban layout. After the 1988 earthquake, Gyumri faced numerous challenges, leading to the gradual degradation of its historical heritage. Key issues included depopulation, a decline in economic activities, and the loss of its role as a regional hub. However, the uniqueness of its urban context, especially its ancient core, offers a potential catalyst for reversing depopulation and reactivating the urban economy, positioning the city as a hub for sustainable development.

This volume presents the results of a consultancy funded by the Asian Development Bank (ADB) and requested by the Urban Development Committee of the Republic of Armenia (UDC) to inform the ongoing urban plans and preserve the historic core of Gyumri.

The guidelines, developed through the consultancy and expanded with additional chapters for context, stem from an in-depth understanding of the city gained through field visits, focus group discussions, ongoing exchanges with the UDC, and consultations with local authorities, stakeholders, and the public.

The final work aims to support the implementation of the planning and regulation for Gyumri and provide a methodological approach for the planning of historic centers that can be replicated in other Armenian cities and similar contexts abroad.

The guidelines are intended to assist local authorities, the general public, and decision-makers in protecting the historic environment and buildings, as well as to serve as a tool for preserving urban identity and driving economic development. Additionally, they offer practical suggestions for residents and professionals involved in the preservation of buildings and open spaces.

Through the Gyumri case study, the volume proposes a working procedure to ensure the preservation of a historic center that has undergone significant transformations over the years while also testing a method that can serve as a reference for other historic towns with similar characteristics, challenges, and potential.

Milan, Italy Mariacristina Giambruno
Milan, Italy Sonia Pistidda
Yerevan, Armenia Nune Petrosyan

Acknowledgments

The consultants are Mariacristina Giambruno and Sonia Pistidda, with the collaboration of Francesca Vigotti and Roberta Mastropirro.

Essential support was provided by Nune Petrosyan, author of Chap. 3 of this volume, vice-chair of UDC, and full professor at the National University of Architecture and Construction of Armenia (NUACA), who, with great passion and expertise, promoted and supervised the study for the city of Gyumri.

We extend our thanks to the Municipal Administration of Gyumri, the students of Architecture at NUACA University who assisted with data collection about the city, the TUMO center in Gyumri that hosted some of the meetings with stakeholders, and Maria Pia Ancora, who oversaw the consultancy activities for ADB.

We also wish to thank the Ambassador and the Italian Embassy in Yerevan, particularly Annarosa Colangelo, who continues to provide information on ongoing promotion and enhancement initiatives in Gyumri. A heartfelt thank you goes to Gohar Hovakimyan and Kristina Hakobyan, who have been following the activities of the working group in Armenia since 2012, and to Gaiané Casnati, through whom we had the opportunity to get to know this country. Finally, special thanks to Antonio Montalto, Honorary Consul of Italy for many years, not only for sharing his deep knowledge of the city but also, and especially, for everything he has done and continues to do for Gyumri and its people. The book is the result of the discussion and mutual collaboration between the authors. In particular, M. Giambruno is responsible for the Chaps. 1 and 4, S. Pistidda for the Chaps. 2 and 5, N. Petrosyan for the Chap. 3. The Chap. 6 is shared by M. Giambruno and S. Pistidda.

Contents

Chapter 1
The Issue of Urban Conservation. Historical Perspective and Contemporary Challenges

Abstract The given chapter traces the most important passages in the history of urban conservation in Europe, focusing particularly on Italy. It begins by recalling the leading figures Camillo Sitte and Charles Buls, who were among the first to defend the ancient parts of cities from the extensive erasure of historic neighborhoods conducted for the sake of hygiene at the end of the nineteenth century. Thereafter, it elaborates on the theory of *diradamento edilizio* ("selective clearing"), which, opposing total demolition, was put forward by Gustavo Giovannoni to sanitize historic centers while preserving their character. With the aim of providing an overview of international debates on the historic city since the second half of the twentieth century, the main international documents are presented and scrutinized in terms of their innovative content. The chapter concludes with an outline of contemporary issues and challenges for the historic city today.

Keywords Historical outline of urban conservation · International documents for the conservation of historic cities · Problems and themes for the historic city in the present day

1.1 Introduction

The urban conservation strategies formulated for Gyumri are certainly derived from the confrontation with potentials and criticalities, socio-economic concerns, problems of urbanism and architecture, as well as contemporary challenges of the Armenian city and historic cities at large. At the same time, these strategies are deeply rooted in the articulate and elaborate debates of the last two centuries, which addressed the so-called "historic city center" and, more generally, the relationship between the existing city and the ever-changing needs of different times.

Accordingly, a brief overview of the underlying theoretical background will be of considerable help to outline the issues, clarify common terminologies, and present specific proposals for the Armenian city, which is considered a stratified palimpsest in its entirety. It should be defended as a whole, not only for its so-called monuments.

M. Giambruno et al., *Heritage-Based Tools for the Sustainable Urban Development of Historic Centers*, PoliMI SpringerBriefs, https://doi.org/10.1007/978-3-031-75259-9_1

Above all, it is a testimony of a culture that is manifested first and foremost in terms of material heritage, and only then in architecture and urbanism.

The considerations that led to the visions and strategies for Gyumri, respecting its historical character and substance, are unthinkable without the centuries-old debates mentioned above.

The overview will cover only the most important passages, which are useful to recall and trace respective logical connections in order to understand the origins of the so-called "historic city" in Europe and Italy, as well as the pioneering voices that defended historic continuity against the 19th-century wave of demolitions related to the needs for comfort, traffic, and hygiene [14, 15]). In this context, the chapter addresses the emerging concept of the city as a stratified palimpsest and the theory of *diradamento edilizio* ("selective clearing"). The former is shaped by different layers of time, where the "minor" everyday built fabric is as important as "monuments" in preserving identity.

The latter, articulated by Charles Buls, implemented in Italy, and gaining international recognition through Giovannoni's work, was, at the time, the only alternative to counter the widespread erasure of historic neighborhoods.

To conclude, the overview offers a critical analysis of international documents concerning the historic city and provides insights into contemporary challenges that now threaten its conservation, much as they did two centuries ago.

This chapter should not be read as a summary of the history of urban conservation in Europe. Instead, it offers a set of insights embedded within a historical perspective that are useful for addressing contemporary issues in urban policies and planning.

1.2 The Nineteenth Century City: Modernization Versus Conservation and the Origins of the "Historic City Center" Issue

The last decades of the nineteenth century were marked by the completion of numerous demolitions of ancient urban cores, following Georges-Eugène Haussmann's Parisian model designed to adapt cities to new hygienic, functional, and mobility needs.

At this time, engineer and urban planner Reinhard Baumeister [5] highlighted in his internationally circulating writings the new relevance of establishing an *umfassender Plan* ("general plan") and municipal regulation policies in light of the latest findings in the field of hygiene.

From that point onward and for a considerable period, architectural legacies of the past were viewed as obstacles to the development of the modern city, to the extent that only a few monuments were preserved by isolating them from their original context. A few years later, in 1889, Camillo Sitte published his book *Der Städtebau nach seinen künstlerischen Grundsätzen*, which marked the first turning point in the relationship between ancient nuclei and the development of the modern city. In this context, the figure of Sitte assumes great relevance since the Austrian scholar is among the first to stand up against the destruction of ancient nuclei.

Sitte's studies are shaped by an idea that will become fundamental for later theoretical developments regarding the existing city: "monuments" should not be viewed as isolated elements removed from their urban context but as integral components of it. In the preface, Sitte clarifies the purpose of his work: "to offer technicians some examples of theoretical deductions, which would serve as a first step towards the great edifice of applied aesthetics and could provide a contribution of experience to the urban planner presenting an allotment plan" [25, p. 15].

The scholar's intention is thus to demonstrate, through studies and examples, both in writing and illustrations, how the organization of ancient cities can be valuable to those planning new neighborhoods. Although this is the primary aim of Sitte's book, the text is also rich in innovative ideas regarding the need to protect ancient urban cores from 19th-century demolitions. The author himself notes that his study wishes to contribute to "saving our old towns, if there is still time, from the increasingly escalating destruction" [25, p. 23]. While he acknowledges that the transformation of ancient urban cores is inevitable, particularly due to hygiene-related issues, he also advocates for a meticulous study of these areas to minimize the necessary demolitions. Furthermore, Sitte engages in a polemic against the isolation of monuments, such as Gothic cathedrals and city gates. He argues that these structures were designed to coexist within their urban environment and are deeply connected to it. While Sitte's vision still favors a reading of the ancient city from a picturesque and idealized vantage point, his message was amplified by the widespread distribution and translation of his book into many languages. It introduced, to contemporary culture, the fundamental idea that city planning must be measured against the reconciliation attempt between the needs of the modern city and the preservation of ancient nuclei.

During the same period, in Belgium, scholar Charles Buls made a crucial contribution to the defense of ancient urban cores through his work. Buls' work had less impact compared to his contemporaries, perhaps due to the excessive contextualization of the examples proposed as case studies, which were entirely taken from Belgian reality [6, 7]. More significantly, his work lacked predefined solutions that could have been readily exported to other contexts. This, although in line with the author's declared intentions, evidently impeded the widespread diffusion of Buls' opinions. In fact, his line of thought is clear and fully comprehensible only in relation to specific cases. When transposed to the purely theoretical realm, his considerations remain largely unintelligible. Charles Buls was born in Brussels in 1837. He was a contemporary of Camillo Sitte and preceded Gustavo Giovannoni by one generation. Mentioning the Belgian scholar alongside the aforementioned well-known figures is no coincidence: Buls' work is very close to Sitte's theories and directly influenced Giovannoni's urban principles.

Buls was mayor of the city of Brussels from 1881 to 1889. In this role, he promoted the preservation of the Place d'Or as an ensemble that could only be protected as a whole, beyond individual monuments. The starting point of Buls' thinking is the conviction that architecture, and consequently, the city represents the set-in-stone identity of the people who produced it. On the contrary, he believed that contemporary urbanism proposed planning models that were in no way suitable for the characteristics of the places to which they were applied.

Among Buls' writings, those of greatest interest for reconstructing the origins of the "question of the existing city", understood as the preservation of its ancient nuclei, are *Esthétique des Villes* (1894) and *L'Isolation des Vieilles Églises* (1910).

In the former, he analyzed the transformations imposed on the city from three different points of view, technical, aesthetic, and archaeological, which gave rise to different solutions (not explicitly elaborated on by the author) in relation to the preservation of the existing urban fabric and the transformations that might be necessary to adapt it to contemporary housing and urban planning standards.

The set of proposed solutions constitutes the theoretical framework of Buls' work, which, in essence, is based on the need to carefully evaluate the interventions to be carried out. These interventions should be minimized to those strictly necessary to improve the living conditions of the inhabitants while preserving the overall layout and perception of the city [23].

Giovannoni's well-known theory of "selective clearing" found an important role model in Buls' ideas [16]. Buls' approach focused on limiting and controlling demolitions while aiming to improve the viability and hygienic standards of old neighborhoods. This involved a continuous mediation between preserving the existing fabric and accommodating the transformations required by urban development. In *L'Isolement*, a publication that followed a lecture by the Belgian scholar in Lübeck in 1908, Buls analyzes the frequent practice of isolating cathedrals.

Although convinced that this practice should be avoided, Buls constructs a methodological framework to evaluate the use of isolations through six points: the importance and style of the building; the historical, artistic, and archaeological value of the constructions adjacent to it; the consequences of the extensive removal of historic neighborhoods on circulation and commercial activities in the area; the views resulting from the demolitions; and, finally, their effect on the spatial perception of the church. As Buls himself states at the beginning of the treatise, he does not align with either the category of "conservators" or "innovators". Instead, he represents a third alternative. In cases of lesser importance, this approach sacrifices parts of the building fabric for the needs of the city's growth. However, it aims to preserve "as much as possible" when the area is significant for the nation's history, serves as a suitable backdrop for "monuments", and aligns with new requirements related to "decorum", "hygiene", and "comfort".

The final part of the overview on theoretical debates about the preservation of the existing city concerns the report written by Alois Riegl in 1903. This report focuses on the "historic center" of Split, in present-day Croatia, specifically referring to the site of the former Diocletian's Palace.

Riegl, whose well-known theories about the notion of "value" had paved the way for contemporary approaches to preserving the built environment, was tasked with deciding whether the archaeological remains of the buildings, despite being more recent but constructed on top, should take precedence in the urban layout.

Riegl's approach essentially applies two fundamental values to the urban scale: the *historischer* and *Alterswert* ("historical and ancient value"), with priority given to the latter as essential for interpreting and understanding more recent constructions. His report concludes with a defense of medieval and modern Split, emphasizing that it is "so inextricably linked to the testimonies of ancient times, with its unparalleled and irreplaceable environmental appeal, requiring the urgent creation of an effective protection law capable of ensuring its integrity beyond merely the primary scientific interest in preserving the ancient remains of the Palace alone" [24, p. 340].

1.3 Debate and Tools for the Defense of Historic Centers in Italy During the Period Between WWI and WWII

Gustavo Giovannoni undoubtedly plays a leading role in both the Italian and international panorama of conservation culture. His prominence is due to the breadth of his published works, the longevity of his career, spanning through wars and profound political changes, the significant influence of his theoretical work on subsequent generations, and his contributions to programmatic documents and protection laws. In Italy, his impact was evident in the conservation practices and legal frameworks that remained active until the end of the twentieth century.

As early as 1913, the year of the publication of his essay *Vecchie città ed edilizia nuova* in the magazine *Nuova Antologia*, Gustavo Giovannoni emerged as a leading figure in the debate on the preservation of ancient cities [16].

Two fundamental concepts that Giovannoni frequently revisits in his work are the shift from focusing on individual monuments to the urban fabric as a whole and the theory of "selective clearing", which opposed the practice of extensively erasing historic neighborhoods. In his work, the scholar analyzes with great clarity the role that old cities should play in relation to new expansions. He emphasizes that these centers are condemned to inevitable destruction due to speculation driven by the impact of unplanned contemporary developments. This includes the concentration of functions in central areas and the resultant increase in economic value, which attracts and polarizes the attention of the real estate market.

Another crucial concept emerging from his writing is the idea of the city as a palimpsest, reflecting the architectural production of successive generations within its urban perimeter. Giovannoni is aware that cities result from consecutive additions from different time periods, with each layer building upon the previous one. This process also includes the transformation of old buildings and environments through the introduction of new styles. The issue, rather, concerns the expressive language of contemporary architecture. Specifically, the "new" architecture has not yet achieved a character and dignity sufficient to be considered a source of enrichment for cities and for the creation of historical stratifications.

In his view, when new buildings are introduced into existing urban nuclei, permitted only in rare cases and when strictly necessary, they should be subordinated to the historic built environment. This can be achieved through the simplicity of architectural lines and respect for existing volumes and colors.

Nearly twenty years later, in 1931, Gustavo Giovannoni revisited and expanded upon what he had outlined in 1913 in the volume *Vecchie città ed edilizia nuova*, which shared the same title as his earlier article. This work became fundamental to the history of "urban conservation".

The theory of "selective clearing" represents a kind of compromise between the integral conservation of the urban built environment and the extensive erasure of the urban fabric, which was still being practiced during that period.

Giovannoni excluded the widening of existing streets, believing it would inevitably lead to the demolition of bordering buildings. Instead, summarizing and simplifying his approach, he proposed creating new streets that would follow the morphology of the existing neighborhood. This approach aimed to enhance the livability of old buildings by removing elements that had increased their height and crowded their inner courtyards. He also suggested making monuments more visible and appreciable through partial demolitions of small structures encroaching on them, implementing green areas to enrich squares, and improving housing hygiene by creating new and enlarged window openings.

While the theory of "selective clearing" can already be found in the work of Charles Buls, Giovannoni introduced the original idea of assigning public administrations the task of managing transformations and guiding private interventions through the realization of exemplary projects.

In this regard, Giovannoni believed that the correct execution of the operation could only be achieved if the urban block were considered as a sort of "minimum unit" for interventions. Giovannoni, therefore, proposed the establishment of consortia to coordinate the works in a unified manner, overcoming the financial interests of individual owners. In terms of implementation, he suggested several innovative and timely measures, including financial incentives, tax breaks, and land credits, while also envisioning a contribution from the public sector to the works [27].

Giovannoni's theory was effectively applied in the plan for "Bergamo Alta", the oldest part of the city, which was developed by Angelini [3]. Conceived between the late 1920s and early 1930s, the plan was implemented between 1936 and 1960, with a hiatus in construction from 1943 to 1950 due first to World War II and then to the challenges of post-war reconstruction (Figs. 1.1, 1.2, and 1.3).

As a product of its time, many aspects of the project adhered to prevailing theories of the "picturesque" and frequently employed the concept of "liberating monuments". The Plan for "Bergamo Alta" thus not only implements interventions related to "selective clearing" to "sanitize" the historic core but also includes urban design projects. These include reconstructing buildings in a simplified yet stylistically coherent manner, creating covered pedestrian pathways, green areas, furnishings and decorations, connecting staircases between different levels, and public buildings. Although the works were completed sixty years ago, the Plan for Bergamo Alta still provides valuable insights for planning within existing urban contexts. Firstly,

Fig. 1.1 Bergamo Alta historic center (*Photo* S. Pistidda, 2024)

Fig. 1.2 Bergamo Alta historic center, Piazza Mercato delle Scarpe (*Photo* S. Pistidda, 2024)

Fig. 1.3 Bergamo Alta historic center, Piazza Padre Reginaldo Giuliani (*Photo* S. Pistidda, 2024)

it relates to the meticulous methods of assessing the existing city, which formed the foundation for every decision regarding the built environment. Each intervention was based on a thorough and precise process of understanding the "Città Alta", with its features meticulously assessed and documented. This comprehensive analysis resulted in a plan that was highly detailed and carefully crafted.

The extensive knowledge of the status quo, combined with the understanding that the success of the plan relied on meticulous attention to detail and stringent frameworks, led Angelini to develop a detailed schedule for the interventions.

Another very interesting feature of the plan is the establishment of a consortium of private property owners, which was made mandatory to manage interventions on an urban scale rather than just individual buildings. Additionally, a planning office was created under Angelini's leadership to support the technical municipal office in the implementation of the plan and to provide advice to private owners.

To assess the frameworks for safeguarding the existing city in Italy and distill what remains useful and relevant today, there are two additional aspects that still need to be considered, setting aside the extensive and complex debate on post-World War II reconstruction [1].

The first aspect concerns the so-called "Charter of Gubbio" [2], an Italian document that can, to some extent, be considered the first "Charter" for historic centers. Arising from the emergency caused by the destruction of historic city centers—first by the World War and then by real estate speculation—the programmatic document resulted from a conference titled *Salvaguardia e risanamento dei centri storico-artistici* (Safeguarding and Sanification of Historic and Artistic Centers) held in Gubbio in 1960. This conference marked the beginning of a new approach to the protection of historic city centers, shifting towards an active policy of safeguarding ancient nuclei. The following year, the National Association of Historic and Artistic Centres (ANCSA), which is still active in Italy today, was founded by a group of scholars who had organized the conference [11].

The "Charter", fully reflective of its historical context, reiterates several concepts that remain relevant today when addressing "historic city centers". Firstly, it rejects the *ripristino* or reconstruction in a historical style, the demolition of historic buildings even if deemed of minimal "architectural value", the thinning out of building stock, and the isolation of "monumental" elements.

Even more interesting are two additional aspects. The first is the need to integrate the most ancient areas into the city's general plans, avoiding the "freezing" of the historic city center and its separation from the contemporary urban fabric of which it is a part. The other aspect is the need to consider the social structure when drawing up plans. This provision is particularly relevant given that gentrification processes, both completed and ongoing, are present in nearly all major cities around the world.

The final missing element in this overview is the "Bologna Plan" of 1973, a *Piano per l'edilizia economico-popolare* (PEEP) ("Plan for Affordable Social Housing"), presented by Pierluigi Cervellati in 1972 during his tenure as a city councilor [8].

The Plan is perhaps the best Italian example of preserving the social fabric and maintaining public housing in historic city centers, an issue that remains relevant today (Figs. 1.4, 1.5, 1.6 and 1.7).

Fig. 1.4 Bologna's historic center, via dell'Inferno. In the foreground, several buildings have undergone a recent urban "regeneration" project and are partly designated for affordable rental housing (*Courtesy* by N. Santopuoli, 2024)

Fig. 1.5 View of Bologna historic center (*Courtesy* by N. Santopuoli, 2024)

Fig. 1.6 Bologna's historic center. The traditional buildings with arcades on the ground floor are designated for commercial spaces, while the residences are located on the upper floors (*Courtesy* by N. Santopuoli, 2024)

Fig. 1.7 Bologna's historic center. View of the porticoes, which were recently added to the World Heritage List (*Courtesy* by F. Ceccarelli, 2024)

Despite the criticisms related to the "restoration" practices proposed in the plan, which involved a *ripristino tipologico* ("typological recovery") of buildings, the urban planning tool was internationally well-regarded for its proposal of affordable and social housing interventions within the historic city center.

A survey conducted in the years leading up to the plan revealed significant deficiencies in the social composition of residents in the historic city center, notably the predominance of elderly residents and pensioners.

During the building refurbishment process, the plan included provisions for the temporary relocation of tenants to buildings constructed specifically for this purpose in the urban voids left by the war (rotational apartments). The spaces in the arcades, which characterize the historic city center of Bologna, were allocated to artisans and commercial activities, preserving the existing ones and ensuring a mix of functions and neighborhood services for residents.

The challenge was to maintain the integrity, at least typologically, of historic buildings while creating residential units that met the required housing standards.

Accommodating the extensive social and cultural services were the convents, which, since their suppression during the Napoleonic era and pre-unification period, had been used as prisons, barracks, and offices.

The Bologna Plan was highly innovative due to its political and social content, and it required considerable perseverance from the city administration to intervene in privately owned buildings. For the establishment of social housing, the possibility of expropriating property owners was introduced. However, this measure was ultimately not used, as an agreement was reached between the public interest and private owners through respective agreements with the municipality.

1.4 Charters, Recommendations, and International Conventions: A Critical Overview

Numerous documents, including conventions, recommendations, and "charters", have been drafted by international bodies and signed by countries worldwide, addressing the protection and conservation of built heritage.

Starting with the first "Charter" created during the First International Congress of Architects and Technicians of Historic Monuments in Athens in 1931—the so-called "Athens Charter"—these documents have evolved over time [17]. Initially addressing architectural heritage as a whole, including protection and general principles for intervention, they have increasingly become more specialized, focusing on specific aspects such as historic gardens, vernacular heritage, and industrial heritage. Other documents began to concentrate on the historic city, addressing both general principles and specific issues related to urban preservation.

The charters resulting from the International Congress of Architects and Technicians of Historic Monuments in 1931 and 1964, known as the "Athens Charter" and the "Venice Charter" [18], do not specifically address the historic city. However, given the significance and international influence of these documents, it is worth noting the few instances where they make indirect references to the oldest parts of the city. In both charters, the focus is primarily on "monuments", with the urban space serving merely as their "environment" that needs to be preserved. This includes safeguarding the immediate surroundings of the monuments, avoiding new constructions, and respecting the "physiognomy of the city" (1931).

The theme of the historic city is more elaborately addressed in the Venice Charter of 1964 than in the Athens Charter. Article 6 states that the "monument", as the central focus of interest, must be preserved in a "setting which is not out of scale". For this reason, it is considered necessary to preserve the "traditional settings" in their mass and color, avoiding the insertion of new buildings or the demolition of existing ones. Although undoubtedly significant, it is important to note that the "Venice Charter" expands the definition of "monument" to include not only the individual structure but also the "rural or urban setting". The article reminds us how the ancient city was still viewed as a picturesque and romanticized place, perceived as being almost timeless and free from more recent "intrusions". The centrality of monuments in relation to the urban fabric is further emphasized in Article 14, where urban spaces (to be considered "monuments") appear merely relevant when framing an allegedly exceptional, "monumental" site.

In the early 1960s, it is evident that the historic city, although recognized as being of "cultural significance", was still somewhat divided into "monumental environments" and the broader historic urban fabric. Conservation and urban planning remained two separate endeavors that were challenging to reconcile.

A little more than a decade later, the issue of the historic city is addressed with greater determination in the "Amsterdam Declaration", also known as the European Charter of Architectural Heritage, adopted by the European Council in October 1975 [20]. The document reinforces the conviction that European heritage encompasses not only the most significant "monuments" but also the "minor" buildings within historic centers and villages, which are closely linked to their built environment or natural context. Additionally, it affirms that historic centers represent a harmonious collective development that fosters social integration. Therefore, renovations in the oldest city districts should be undertaken without altering the social composition of the residents. Contemporary with the Amsterdam Declaration, the Bologna Plan addressed these issues and, in terms of social aspects, transformed them into an instrument of considerable quality.

The Declaration overcomes the dichotomy between planning and conservation that was still evident in the Venice Charter. It asserts that planning must integrate the requirements of architectural heritage conservation, fostering a continuous dialogue between conservationists and planners.

In the same year, ICOMOS adopted the Resolution of the International Symposium on the Conservation of Smaller Towns during its 4th General Assembly [19]. The document, while focused on the specific field of urban conservation and the challenges of protecting small towns, includes some passages of broader relevance.

In particular, it addresses the issue of depopulation (a significant concern at the time), the need for small towns to be preserved in an integrated manner with regional and urban policies, and the role of tourism. The document highlights the positive potential of tourism for revitalizing marginalized places while also cautioning against the potential threats it poses to the appearance and structure of the city.

Forty years after the adoption of the document, the debate on the role of tourism in historic cities remains highly relevant and contentious, particularly concerning the preservation of historic urban environments and the impact on both physical integrity and social diversity.

A decade later, the ICOMOS General Assembly adopted the Charter for the Conservation of Historic Towns and Urban Areas (Washington Charter, 1987), presenting it as a specific supplement to the Venice Charter concerning the conservation of historic towns. The document addresses a range of issues related to the oldest parts of cities, including traffic and parking problems, natural disasters, and the specialized training required for various conservation skills. An important point emphasized by the Charter is the need for the conservation of the historic city to be integrated into the economic and social policies of a specific area, as well as being embedded in urban and regional planning at every level [21].

The historic city must be considered and preserved in all its aspects, including the urban form defined by lots and streets, the relationship between buildings and open spaces, and the connection with surrounding natural or man-made areas.

This open and inclusive view of the many aspects that constitute the existing city and the emphasized need for their preservation in the Charter contrasts sharply with the recommendations regarding buildings, which focus solely on the preservation of formal aspects. The Washington Charter also outlines the characteristics that an urban conservation plan should address. It should be preceded by multidisciplinary studies, regulate the relationship between the historic part and the city as a whole, and define which buildings to preserve and which to demolish (exceptionally). The primary goal, however, should be to improve housing conditions in residential buildings.

Finally, two other aspects presented in the document are worth mentioning. Firstly, the importance is given to involving the resident population, starting with the youngest (school children). Secondly, the document addresses the theme of integrating new elements into existing built environments, a topic that remains highly relevant today. It goes a step further than the Venice Charter by allowing the addition of new elements, viewing it as an opportunity to enrich certain parts of the historic city. However, as a programmatic and general document, it can only stipulate that these new elements should be "in harmony" with the existing fabric and respect the scale and dimensions of the lots.

2011 was an exceptionally productive year for international documents concerning the preservation of historic cities. In November 2011, UNESCO adopted the *Recommendation on the Historic Urban Landscape* on the 10th [26], followed by ICOMOS's adoption of *The Valletta Principles for the Safeguarding and Management of Historic Cities, Towns, and Urban Areas* on the 28th. The second document is, to some extent, an update of the Washington Charter in terms of content and proposed definitions [22].

An interesting aspect of these documents is their broadening of the definition of "historical cities and urban areas". They consider these areas as integral parts of contemporary life and essential bases for urban and territorial development. This perspective acknowledges both tangible and intangible elements of urban heritage. As a result, any urban area with cultural significance is eligible for listing, irrespective of its stage in urban development.

The Valletta Charter also contains a series of proposals and strategies for the preservation of historic cities. It defines, for one, the elements that should be conserved: the integrity and authenticity of the historic city, the urban form, the shape of buildings, the relationship between the city and its surroundings, the acquired functions, the cultural and technical traditions, the spirit of the place, as well as the non-renewable resources. Moreover, it states that new functions should not compromise traditional activities, reiterates that contemporary architecture should fit into the ancient urban fabric while respecting its character, and emphasizes that services should consider the residents' needs.

The document identifies two instruments for the historic city: the conservation plan and the management plan. This subdivision suggests that the Valletta Charter is addressed not so much to urban built environments in general, but specifically to historic towns on the World Heritage List. The Recommendation on the Historic Urban Landscape is dedicated to the topics above and includes some useful points on more general aspects. The first concerns the definition of Historic Urban Landscape (HUL), which is described as an area resulting from the process of urban stratification that goes beyond the concept of the historic city center.

"The historic urban landscape approach is aimed at preserving the quality of the human environment, enhancing the productive and sustainable use of urban spaces while recognizing their dynamic character, and promoting social and functional diversity. It integrates the goals of urban heritage conservation and those of social and economic development. It is rooted in a balanced and sustainable relationship between the urban and natural environment, between the needs of present and future generations and the legacy from the past." [26, art. 11].

Another important point is the recognized need to integrate the protection of HULs into large-scale policies and planning, as well as into the development agendas of individual states. It also suggests, as a necessary tool, the civic involvement of as many stakeholders as possible, the knowledge and monitoring of changes, and the development of adequate regulatory systems and financial support measures.

This brief (and thus partial) overview of documents concerning the historic city, ranging from 1964 to 2011, shows how numerous issues related to the oldest parts of cities have been addressed by the major international bodies dealing with the conservation of architectural heritage.

The definition of the historic center was gradually extended to encompass the entire city as a whole and its stratified layers of tangible and intangible assets—a *bene culturale* ("cultural heritage") in itself. The addressed issues include the introduction of new elements into the existing urban fabric, tourism as both a resource and a threat, the involvement of local communities and their needs, and the challenge of maintaining the resident population and traditional activities in the face of gentrification. Even after more than a hundred years of debate and numerous recent international documents, historic cities continue to face many challenges today [9, 10, 12].

1.5 Contemporary Challenges for Urban Conservation

Much like living organisms, the urban fabric of historic cities is the result of profound transformations over time, shaped by the individuals who have lived there. Attempting to freeze the historic city at a given moment in time would not only be contrary to its nature but also impossible. Like all living organisms, it changes, and within these changes lie its vitality and, to some extent, the possibility of its physical preservation (Fig. 1.8). However, these observations must be accompanied by two notes. The first concerns overcoming the concept of the "historic city center".

When we speak of a historic city, or more precisely an "existing city", we are referring not just to the oldest part but to the entire urban fabric, which extends back at least to the 1950s, when the culture of urban planning and architecture underwent significant changes in both form and approach. Even this temporal constraint could be surpassed and brought closer to the present, given the increasingly rapid changes imposed by contemporary times.

The second concerns how necessary transformations are implemented. While previously, the continuity of the past was considered fundamental, and there was an effort to reproduce it, today, we are well aware that it is not possible to go back: what is lost is lost forever. In other words, the demolition of entire urban areas should be the subject of extensive reflection, study, and investigation, far beyond what is normally carried out.

Fig. 1.8 The block of the Rossija Hotel in Moscow after the demolitions. On the panels that make up the construction fence, there are images of the buildings before the construction of the hotel (*Photo* M. Giambruno, 2013)

The historic city must, therefore, face many challenges today, balancing the need to maintain its role as the vibrant core of the contemporary city while also preserving its character and substance [4, 13].

The extensive transformation of urban areas affects many cities in so-called developed countries. This is particularly true for areas that have been gradually abandoned, such as former peripheral industrial sites now located centrally or semi-centrally, barracks, railway yards, and similar sites.

Vast areas that had been unused or underused for years have been demolished in recent decades to make way for multi-story residential buildings, shopping centers, and spaces for the tertiary sector.

These new areas, which are certainly impactful and somewhat attractive, especially for tourists visiting the city as in the case of Milan, have, however, led to the total disappearance of what had been established on the site for years (Figs. 1.9, 1.10, 1.11 and 1.12). As previously mentioned, the goal is not to stop transformations but to ensure that plans for these areas adequately account for permanence, giving rise to projects that enable a dialogue between the new and the old.

Instead, the prevailing idea is that stopping the consumption of more land does not equate to preserving and reusing what already exists but rather to demolishing it, "recycling", and constructing something new. Although this new construction may be more sustainable in terms of energy consumption, it involves the waste of non-renewable resources through the production of new materials and the disposal of the old ones resulting from demolition.

Another issue affecting cities in advanced countries is tourism, which in many places has become mass tourism (Fig. 1.13). As some international documents adopted since the 1970s have indicated, while tourism is certainly a resource, it can also become a threat, leading to the depletion of cultural heritage and altering the social composition of the population.

Gentrification on one hand and hyper-tourism on the other. Numerous residential buildings have been converted into short-term rental properties, leading to a rise in

Fig. 1.9 The area of "Porta Nuova" with its skyscrapers was built in the "Isola" neighborhood in Milan between 2005 and 2015, radically transforming a part of the historic city (*Photo* M. Giambruno, 2024)

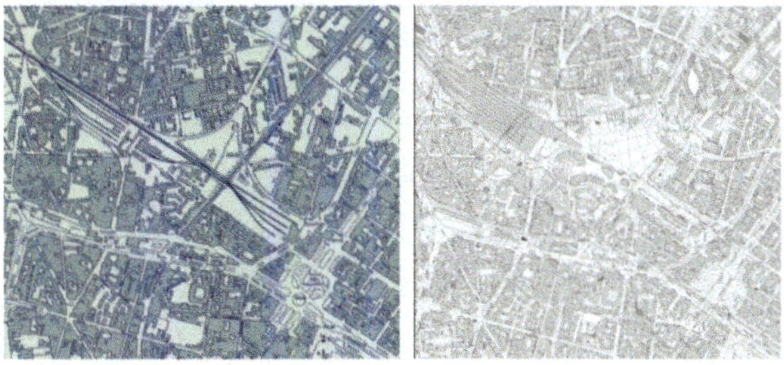

Fig. 1.10 The area of the "Isola" district and the cargo terminal in the 1956 and 2020 maps (Geoportale Municipality of Milan, historical maps, https://geoportale.comune.milano.it/sit/, accessed 31.07.2024)

the real estate market and creating challenges for residents in historic cities, which are increasingly losing neighborhood services in favor of those catering to tourists.

With a law coming into force on September 5, 2023, New York is trying to limit these phenomena that ultimately drain the central parts of the city of permanent residents and profoundly change the economic and social structure. In the cities of the so-called developing countries, the dynamics are even more complex and, in some ways, even less controllable.

There are two opposing trends. In some contexts, the historical part of the city is the most deteriorated, lacking primary infrastructure and inhabited by the less affluent population (Figs. 1.14 and 1.15). In others, particularly in capitals, the process of

Fig. 1.11 The three skyscrapers (Isozaki, Hadid, Libeskind) built in the former trade fair ("La Fiera") in Milan (1923–1990) (*Photo* M. Giambruno, 2020)

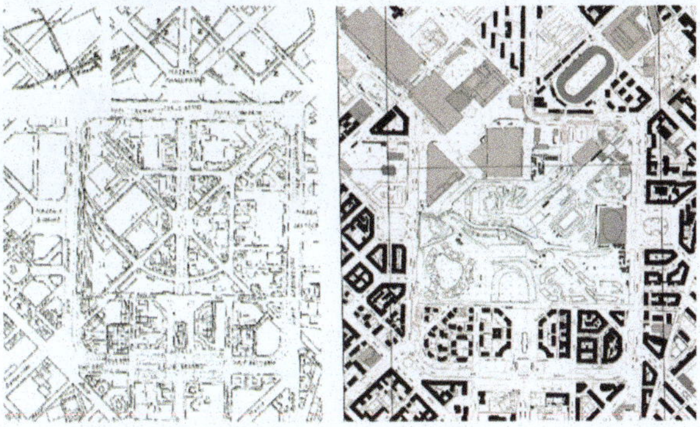

Fig. 1.12 The area of the former trade fair ("La Fiera") in Milan in the 1930 map (left), and 2020 (right) (Geoportale Municipality of Milan, historical maps, https://geoportale.comune.milano.it/sit/, accessed 31.07.2024)

gentrification has led to the expulsion of economically weaker groups who are unable to afford to rent or buy property in the central parts.

For both scenarios, the historical urban fabric is paying the price. In the first case, the area is abandoned or underused without interventions, not even receiving the slightest maintenance, due to the difficult economic conditions of those who live there. This results in serious decay or even the collapse of buildings. In the second, older low-density buildings are replaced by new multi-storey ones in the most ancient part in order to exploit the new land value induced by the market (Fig. 1.16).

Fig. 1.13 Venezia, Piazza San Marco (*Photo* M. Giambruno, 2014)

Fig. 1.14 The Walled City of Multan, Pakistan, the conditions of the built heritage (*Photo* M. Giambruno, 2012)

Fig. 1.15 Lahore city, Pakistan, the densification of the buildings (*Photo* M. Giambruno, 2012)

Fig. 1.16 An intervention in a historic building in the center of Yerevan has resulted in the preservation of only the main façade of the old structure. This façade now serves as the base for a new multi-story construction (*Photo* M. Giambruno, 2022)

This brief overview of contemporary issues for the historic city in different contexts serves to understand how the key points of the centenary debate and international documents can be seized to develop policies and planning tools that are able to safeguard ancient parts of the city, reconciling urbanism and conservation, transformation, and conservation.

References

1. AA VV (2009) Mirror of modernity. The post-war revolution in urban conservation. Docomomo e-proceedings II
2. ANCSA (1960) Carta di Gubbio. https://www.ancsa.org/la-storia-e-larchivio/la-prima-carta-di-gubbio-1960/. Accessed 18 Aug 2024
3. Angelini L (1963) I lavori compiuti per il piano di risanamento di Bergamo alta. Conti, Bergamo
4. Appleyard D (ed) (1977) Urban conservation in Europe and America: planning, conflict and participation in the inner city. In: Conference proceedings, Tipografia Olimpica, Roma
5. Baumeister R (1876) Stadtserweiterungen in technischer baupolizeilicher und wirtschaftlicher Beziehung. Berlino
6. Buls C (1894) L'estetique des villes. Van Oest, Bruxelles
7. Buls C (1910) L'isolament des veilles Églises. Van Oest, Bruxelles
8. Cervellati P, Scannavini R, De Angelis C (1977) La nuova cultura della città: la salvaguardia dei centri storici, la riappropriazione sociale degli organismi urbani e l'analisi dello sviluppo territoriale nell'esperienza di Bologna. Mondadori, Milano
9. Cody J, Siravo F (eds) (2019) Historic cities: issues in urban conservation. The Getty Conservation Institute, Los Angeles
10. Cohen N (1999) Urban conservation. MIT Press, Cambridge (Mass)
11. Di Biase C (1990) 30 anni ANCSA. 1960–1990. Tipografia Cordani, Milano
12. Di Biase C (ed) (1989) Nuova complessità e progetto per la città esistente. Franco Angeli, Milano
13. Fiorani D (2019) Il futuro dei centri storici. Digitalizzazione e strategia conservativa. Edizioni Quasar, Roma
14. Giambruno M (2002) Verso la dimensione urbana della conservazione. Alinea, Firenze
15. Giambruno M (ed) (2007) Per una storia del restauro urbano. Cittàstudi Edizioni, Milano
16. Giovannoni G (1931) Vecchie città ed edilizia nuova. Torino, UTET
17. ICOMOS (1931) The Athens charter for the restoration of historic monuments. https://www.icomos.org/en/resources/charters-and-texts/179-articles-en-francais/ressources/charters-and-standards/167-the-athens-charter-for-the-restoration-of-historic-monuments. Accessed 20 Aug 2024
18. ICOMOS (1964) The Venice charter. https://www.icomos.org/en/participer/179-articles-en-francais/ressources/charters-and-standards/157-thevenice-charter. Accessed 20 Aug 2024
19. ICOMOS (1975) Resolutions of the international symposium on the conservation of smaller historic towns, at the 4th ICOMOS General Assembly. https://www.icomos.org/en/resources/charters-and-texts/180-articles-en-francais/chartes-et-normes/384-resolutions-of-the-international-symposium-on-the-conservation-of-smaller-historic-towns-at-the-4th-icomos-general-assembly. Accessed 20 Aug 2024
20. ICOMOS (1975) The Declaration of Amsterdam. https://www.icomos.org/en/and/169-the-declaration-of-amsterdam. Accessed 20 Aug 2024
21. ICOMOS (1987) Charter for the conservation of historic towns and urban areas (Washington Charter). https://www.icomos.org/images/DOCUMENTS/Charters/towns_e.pdf. Accessed 20 Aug 2024

22. ICOMOS (2011) The valletta principles for the safeguarding and management of historic cities, towns and urban areas. https://civvih.icomos.org/wp-content/uploads/2022/03/Valletta-Principles-GA-_EN_FR_28_11_2011.pdf. Accessed 22 Aug 2024
23. Naretto M (2016) Charles Buls et la restauration. Anthologie critique. FrancoAngeli, Milano
24. Riegl A (1903) Rapporto su una ricerca per la valutazione dell'interesse verso i monumenti medievali e moderni all'interno del palazzo di Diocleziano a Spalato, condotta per incarico della Presidenza della I.R. Commissione centrale. In: Scarrocchia S (1995) Alois Riegl: teoria e prassi della conservazione dei monumenti. Clueb, Bologna
25. Sitte C (1889) Der Städtebau nach seinen Künstlerischen Grundsätzen; Carl Graeser W tr. It. (1980) L'arte di costruire la città. L'urbanistica secondo i suoi fondamenti artistici. Jaka Book, Milano
26. UNESCO (2011) Recommendation on the historic urban landscape. https://whc.unesco.org/en/hul/. Accessed 22 Aug 2024
27. Zucconi G (1997) Gustavo Giovannoni. Dal capitello alla città. Jaca Book, Milano

Chapter 2
Heritage-Based Tools for the Existing Cities: A Methodological Issue

Abstract The debate on sustainable development often overlooks urban issues and Built Heritage. However, historic centers are the areas where global challenges become evident in all their complexity. A shift in perspective is both urgent and necessary. It involves viewing urban cultural heritage and its communities not just as a problem but as part of the solution. Integrating historic preservation into development plans assigns an active role to this heritage in project strategies. This chapter encourages reflection on the importance and prioritization of this legacy by highlighting key challenges and introducing methodological tools for addressing them.

Keywords Urban sustainable development · Heritage and global challenges · Climate change · Minimum quality in planning tools

2.1 Urban Sustainable Development and Global Challenges. Promoting the Art of "Care"

The Brundtland Report of 1987 defines sustainable development as "development that meets the needs of the present without compromising the ability of future generations to meet their own needs" [8]. While it identifies numerous factors involved in sustainability, the report does not explicitly recognize built heritage, and so the urban issue, as a cultural contribution to development.

Nevertheless, historic centers and their built heritage can play a crucial role in addressing contemporary global challenges, especially in terms of urban and territorial sustainability. Historic centers can act paradoxically as both drivers and constraints of economic development.

They represent, by definition, a transdisciplinary territory that encompasses various disciplines and fields, serving as an area of complexity, historically debated between the needs of preserving the historic structure (which includes not only buildings but also urban patterns, the relationship between buildings and empty spaces, and green areas) and those of "planning for transformation". Here, social, economic,

M. Giambruno et al., *Heritage-Based Tools for the Sustainable Urban Development of Historic Centers*, PoliMI SpringerBriefs, https://doi.org/10.1007/978-3-031-75259-9_2

and material aspects intertwine within a dynamic structure, transcending the physical characteristics and embracing elements such as the use of spaces and intangible aspects. Cities in emerging countries are experiencing similar issues to those in more developed contexts: rapid urbanization, climate change, resource depletion, gentrification, mass tourism, and standardized commercial activities.

According to the latest UNESCO report on Climate Action for World Heritage, "Climate change has become one of the most significant threats to World Heritage, impacting the Outstanding Universal Value (OUV), including integrity and authenticity, of many properties, as well as the economic and social development and quality of life of communities connected with World Heritage properties" [7].

This document builds upon earlier reflections on the topic. The discussion, started by the World Heritage Committee in 2005, led to two reports in Vilnius in 2006: *Predicting and Managing the Effects of Climate Change on World Heritage and Strategy to assist States Parties to implement appropriate management responses* [1]. This discussion resulted in the 2008 *Policy Document on the Impacts of Climate Change on World Heritage Properties* [4]. Since the 2007 policy document, numerous reports have highlighted heritage at risk due to climate change.

The rapid evolution of climate threats has led to the understanding that some effects, while devastating, are irreversible, affecting not only natural ecosystems but also threatening human lives.

"Cultural World Heritage properties, represented by cultural landscapes, historic cities, archaeological sites, and vernacular architecture also demonstrate various locally developed strategies for mitigating climate change through energy-efficient built forms and sustainable use of local resources" [7, p. 3].

The urgent need, as outlined by the report, is to "making World Heritage properties carbon neutral, as much as possible, and more resilient and better adapted to a changing climate, while safeguarding their Outstanding Universal Value" [7, p. 3].

Built Heritage has frequently demonstrated intrinsic resistance to external threats. As noted in the ICOMOS [3] report, cultural heritage contains elements of resilience to climate change threats, found in traditional techniques such as self-modification in response to climatic conditions (e.g., adaptations to wind and floods, like shading devices and verandas). The practice of reusing resources from a circular economy perspective is embedded in the traditions of some cultural systems and ancient land-use practices. Thus, cultural heritage can serve as a repository of good practices to enhance resilience through a place-based approach, already possessing inherent natural, cultural, and social resilience.

Urban cultural heritage can, therefore, be part of the solution to climate change, becoming exemplary in embracing the transformation the world needs, with communities playing a key role in its preservation. However, to actively contribute to this change, heritage must be integrated into a systemic vision, involving collective actions and engaging the international community.

Assigning cultural heritage a priority and an active role in guiding project strategies implies incorporating historic preservation into development plans.

The 2023 document seeks to provide an outcome-oriented policy framework to support these changes. The guiding principles aim to minimize risks and increase adaptive capacity through a precautionary approach. Article 23 emphasizes the importance of knowledge as the foundation for all actions: "Use the best available knowledge, generated through disciplinary, interdisciplinary, and transdisciplinary processes, including input from scientists, researchers, site managers, Indigenous Peoples, and local communities" [7, p. 6].

A set of climate action goals for world heritage towards 2030 is established, summarized in key areas: risk assessment, adaptation, mitigation (including passive measures to reduce energy consumption, life cycle assessment for selecting materials, and promoting maintenance practices), knowledge, capacity building, and awareness.

The adoption of the 2030 Agenda in 2015, which identified 17 Sustainable Development Goals (SDGs) [9], reaffirmed the lack of explicit connection with cultural topics. Cultural heritage is mentioned only in Goal 11, "Sustainable Cities and Communities," with Target 11.4 stating: "Protect and safeguard the world's cultural and natural heritage".

In the same year, the World Heritage Committee adopted in Paris the "Policy for the Integration of a Sustainable Development Perspective" into the Processes of the World Heritage Convention [6], aiming to connect Sustainable Development Goals with the climate change issues. The policy recognizes that:

"In the face of increasing disaster risks and the impact of climate change, States Parties should recognize that World Heritage represents both an asset to be protected and a resource to strengthen the ability of communities and their properties to resist, absorb, and recover" [6].

Urban sustainable development is becoming an increasingly urgent goal. Integrating historic preservation into urban planning tools not only honors the past by maintaining cultural identity but also aligns with global sustainability goals and contemporary needs.

2.2 The Widespread Heritage: Facing the Enlargement of Heritage Meanings

Starting from the mid-nineteenth century, major European cities faced the challenges brought about by the demographic growth in post-industrial society, leading to new urban transformation needs. The main debate centered around the dichotomy between preserving historical identity and meeting the demands of change, a theme central to all significant transformations in European cities.

Within these dynamics, the "widespread heritage"—consisting of lesser-known and more fragmented buildings, often referred to as minor architecture—is generally considered expendable. While there is a tendency to protect more monumental buildings through "isolation" interventions, minor buildings undergo a process of marginalization.

This widespread heritage, more vulnerable to abandonment, spontaneous decay, or demolition, thus reveals its inherent fragility.

Despite the crucial role that this minor architecture plays in defining urban morphology, it lacks visibility and appeal compared to more significant monuments. As a result, it often lacks adequate protection, and its conservation and transmission to future generations depend largely on the sensitivity of individuals or communities.

Policies for historic centers, especially in emerging countries looking to rapid development models, tend to encourage investments and promote private initiative, boosting productive activities (Fig. 2.1).

In the absence of clear rules, the potential direct consequences on the territory are evident. Parts of historic cities continue to be demolished to make way for a new, increasingly global, and homogenized concept of the city, with the sometimes-short-sighted aim of providing immediate and pragmatic responses to economic contingencies and pressures (Fig. 2.2). Planning often leans towards procedural simplification, making it more difficult to control the quality of the work and, consequently, to protect the historic center. Terms like "regeneration", "revitalization" and "valorization" are often used in relation to intervention methods, where the built heritage is often the one to suffer. What then can be said when the widespread heritage of historic centers belongs to an uncomfortable past that many would rather condemn to oblivion?

Fig. 2.1 One of the main streets of the Gyumri historic center with the recent transformations (*Photo* S. Pistidda, 2022)

Fig. 2.2 Widespread heritage in Gyumri (*Photo* S. Pistidda, 2022)

But "these centers are defined precisely by their historical nature, and the loss of historical quality leads to the loss of the very identity of the places" [2].

As Piccinato highlighted, "the stratification and interpenetration of different eras are the very substance of the urban" [2], promoting the recognition of a value-generating function even for the widespread heritage.

In the theoretical debate, the Venice Charter of 1964 marks the expansion of the concept of "monument" to the "urban environment", while the Amsterdam Charter of 1975, with its idea of "integrated conservation", is the first international tool to refer to the historic city as "urban heritage". The Washington Charter of 1987, on the other hand, begins to outline operational guidelines, emphasizing the need for multidisciplinary preliminary studies, promoting participation, and conserving the defining elements of the historic center.

Starting in 1978, UNESCO included historic centers in the World Heritage List, and in 2011, it defined recommendations for historic urban landscapes. The document begins by recognizing historic urban areas as "the most abundant and diverse manifestations of our common cultural heritage, shaped by generations and constituting a key testimony to humankind's endeavors and aspirations through space and time" [5].

Urban areas today are the primary locations for growth and life, offering economic and educational opportunities, but they are also the most threatened by uncontrolled development. The cultural value of a historic fabric is not inherently quantifiable, as it consists of factors such as identity recognition, memory from stratifications, and other intangible components. Acknowledging the important role of historic areas in modern societies, the document emphasizes the need for a comprehensive and integrated approach to their conservation and management. It outlines tools that must be integrated into every development program across specific areas: civic engagement, knowledge and planning, regulatory systems, and financial tools.

In Armenia, there are only three UNESCO World Heritage Sites (Monasteries of Haghpat and Sanahin, Cathedral and Churches of Echmiatsin and the Archaeological Site of Zvartnots, Monastery of Geghard and the Upper Azat Valley) and five candidate sites (the archaeological site of the city of Dvin, the Basilica and archaeological site of Yererouk, the Monastery of Noravank and the Upper Amaghou Valley, the Monasteries of Tayev and Tatevi Anapat and the adjacent areas of the Vorotan Valley, and the Vishaps and the cultural landscape of Tirinkatar).

None of the historic centers appear among these sites. Yet, several Armenian historic centers exhibit unique characteristics, irreplaceability, and a richness of cultural elements and stratifications. The risk of loss for this heritage is significant, and there is an urgent need to map the geography of territorial assets to initiate protection efforts.

On one hand, there is an international awareness that the concept of heritage is not limited to exceptional and "monumental" episodes. However, the reality of protection practices is quite different and faces several issues. The foremost issue is the very definition of what constitutes a historic center and its delineation. What should be the subject of protection? Some places also have recent histories; for example, Gyumri was partially rebuilt after the earthquake. In these cases, defining the temporal threshold and the subject of protection becomes even more complex.

The second problem concerns the territorial extent, including the quantitative scope, and the pervasiveness of this heritage, which make it impossible to consider specific protection procedures. Equally important is the issue of ownership, which is mostly private, making it difficult to impose protection and preservation measures. Moreover, public administrations, especially in emerging countries with even more limited staff, would not be able to manage such a large volume of work.

It is, therefore, necessary to think of more agile and manageable tools within urban planning policies, which place urban capital at the center as a resource to be protected and also actively integrated into a virtuous cycle of transformation. This allows it to participate actively in transformation dynamics rather than being a passive subject. The next section begins to outline some initial methodological reflections on incorporating conservation issues into planning and regulations.

2.3 Inserting Conservation Issues in Planning and Regulations. Criticalities and Methodological Issues

To address the issue of historic centers, it is necessary to develop new methodological strategies that can ensure the survival of historic cities while respecting their specific identity. These strategies must also enable a conscious and sustainable transformation, balancing the need to preserve cultural heritage with the contemporary requirements of development and modernization.

Knowledge, as has been repeatedly emphasized, is the fundamental basis for any intervention. However, in this delicate phase of listening and gathering, it is important not only to consider the nature of the information but also how it is collected, organized, and systematized, and consequently, how it can create new data interactions. The availability of an integrated and coherent analytical method is a necessary condition for enabling interactions between the various interpretations of the territory. Digitalization and informatization, therefore, form the foundation for preserving the vast wealth of information that represents an essential prerequisite for the project.

Even a design phase that might seem straightforward is not without its problems. Preliminary analytical knowledge is often viewed as a "time-consuming" operation, to the detriment of the project, and therefore is something to which limited time and resources are allocated. Additionally, the knowledge phase suffers from methodological issues arising from the difficulty of delineating the boundaries of the historic center and identifying its components, as well as from the temporal definition of what is considered "historic". Up to which period? Before or after which historical threshold?

Furthermore, if the debate in Italy has highlighted a lack of convergence between "urban planning culture and conservation culture" [2, p. 34], it is easy to imagine what might happen in contexts where this debate has not even started.

Even in operational terms, the globalized production system favors large companies at the expense of small artisanal firms, leading to the loss of specialized craftsmanship skills.

What can be done? If widespread heritage is difficult to protect due to its extent, heterogeneity, and complexity, planning tools could play an active role if they incorporate minimum quality requirements for projects in historic contexts. The goal would be the active protection of this heritage as an element that enhances the quality of life for residents. Naturally, the implications of this approach are significant: first and foremost, it involves defining the entities responsible for drafting these requirements. This shifts the discussion to the disciplinary figures that should be involved in drafting these requirements, who must necessarily have at least a basic education in the fields of heritage conservation and protection. This issue is directly connected to the topic of university education.

To act promptly, it is also necessary to have a clear mapping of the extent and characteristics of such heritage. For this purpose, an inventory that encompasses not only monumental heritage but also widespread heritage at the urban scale—identifying its characteristics, state of conservation, elements of interest to be preserved, etc.—would be essential (Fig. 2.3). However, it is crucial to avoid the mistake of creating catalogs that merely serve as references for replicating elements identically.

When moving to the operational level, the issues are numerous and of different nature. Too often, however, planning tools at all scales treat historic cores in the same way as other parts of the city, establishing uniform rules that do not account for the heterogeneity of historical elements and are not supported by adequate and in-depth research (once again, the knowledge gap!). Introducing a series of minimum

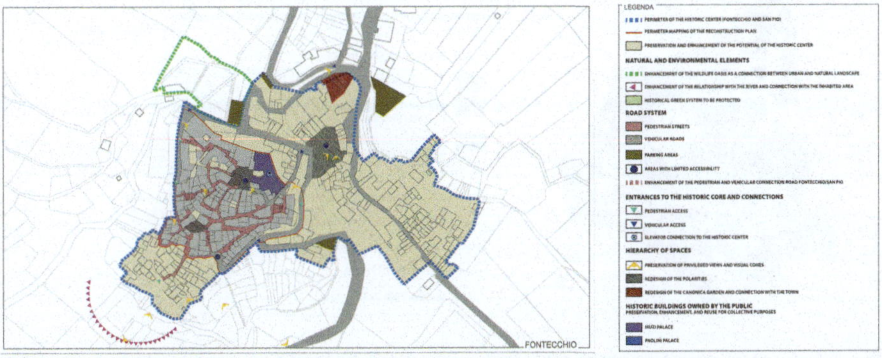

Fig. 2.3 General strategies of the plan. The issues include elements such as accessibility, entrances and connections, natural and environmental elements, and spatial hierarchy. Fontecchio (L'Aquila, Italy), Reconstruction Plan, Politecnico di Milano (© M. Boriani, M. Giambruno, F. Maroldi, R. Simonelli, R. M. Rombolà, S. Pistidda, F. Sebastianutti, S. Lucca, G. Ramponi)

requirements for the presentation of building applications could be a first step towards better control of interventions from a conservation perspective.

Another critical issue concerns the lack of a clear and standardized methodological approach for identifying a historic center, defining its boundaries, and understanding its various levels of historical complexity. Additionally, there is the matter of terminological definitions: what exactly constitutes a historic center? What defines an urban nucleus? Therefore, the primary need would be to create a common glossary to ensure that when certain terms are used, there is alignment on their meanings.

The lack of guidelines in drafting the knowledge project is also reflected in the design directives, which are often generic and do not account for the diversity of pre-existing conditions within an urban center. Planning tools frequently lack specific indications on permissible and impermissible transformation interventions, tailored to the differences and specificities of various building categories (Fig. 2.4).

As a result, responsibility is often left to individual municipalities, which adopt different tools based on the needs of the moment.

At a methodological level, flexible and adaptable tools are needed to integrate with existing planning instruments. The development of a comprehensive, structured, and shared knowledge base is the starting point, not merely as a collection of appendices but as a "project" that facilitates multiscalar comparisons between different data sets, both homogeneous and heterogeneous, and allows for cross-sectional readings from multiple perspectives. This knowledge base should be continually updated, expandable, and overlayable, enabling the creation of differentiated thematic layers.

Data must be organized across multiple levels of information and interwoven. It is also crucial to determine, depending on the purpose, how to represent the existing conditions, their state of preservation, and their relationships with both tangible and intangible elements of the city (such as social, economic, and demographic aspects).

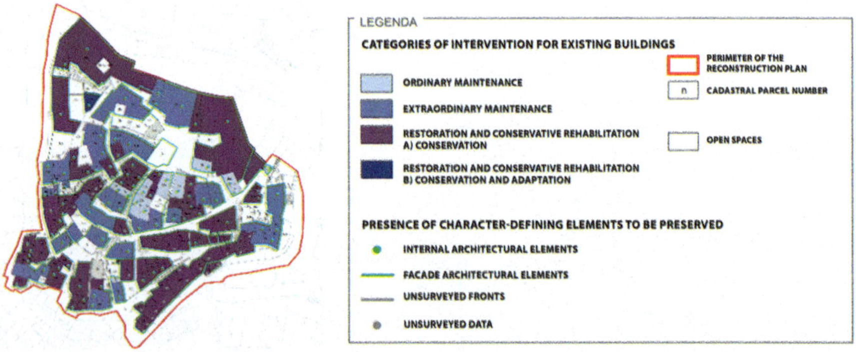

Fig. 2.4 General strategies of the plan, interventions for the built environment. The issues include different categories of interventions from maintenance to conservation. Fontecchio (L'Aquila, Italy), Reconstruction Plan, Politecnico di Milano (© M. Boriani, M. Giambruno, F. Maroldi, R. Simonelli, R. M. Rombolà, S. Pistidda, F. Sebastianutti, S. Lucca, G. Ramponi)

Geographic Information Systems (GIS), including recent developments in artificial intelligence, offer intriguing possibilities for storing large amounts of data and building networks of relationships between them [2].

The work done on the city of Gyumri aims to introduce minimum quality requirements into the methodological approach for reading the historic center, which is an inseparable prerequisite for subsequent effective project actions.

2.4 Plan Drafting Requirements. Defining a Strategic Vision for the Historic Centers

Revitalizing and preventing the abandonment of historic parts of a city requires multiple, coordinated strategies. The historic city is a complex reality where historic and contemporary elements are closely intertwined; old buildings interact with outdoor spaces and the urban layout. Therefore, strategies must be designed for the city as a whole rather than treating the historic center and the contemporary city as two separate entities (e.g., the historic area as merely a tourist attraction without necessary services for residents and the contemporary part solely as a place for locals).

The historic center should serve as a vibrant core of the city, maintaining a diverse mix of functions—residential, tourist, administrative, and commercial—that are closely aligned with the characteristics of its spaces and historic buildings.

The gentrification of the historic center, a phenomenon that has affected many European cities, should be avoided. It is crucial to maintain a balance among the various social backgrounds of the population to preserve a vibrant and diverse urban environment. The main characteristics of a historic city are embodied in its buildings and urban form, which have evolved over time. Therefore, any strategy should aim to preserve and enhance this historical integrity, avoiding demolitions, alterations to the urban layout, and reductions in historic fabric.

Interventions must be coordinated both in their theoretical approach and practical implementation. Municipalities and public bodies should manage this process through master plans, detailed plans, and guidelines. Given the available resources and the legal framework, establishing a dedicated office for the historic center could be beneficial in steering the process effectively.

This approach would facilitate the management of both small projects undertaken by private citizens and larger initiatives funded by national and international investors. While attracting private investors can significantly contribute to the revitalization of the historic center, it is essential to prevent entire areas from being dominated by such interventions. These efforts must, in any case, be guided by public administration to ensure balanced development.

To summarize and outline effective strategies for historic centers, the following steps are necessary:

- **Analyze the role and position of the city within the country**: This includes a detailed examination of demographic trends, economic conditions, current and disused productive activities, social composition, population age, and education levels, transportation connections (both domestic and international), and existing public services.
- **Assess the historic center**: Study the specific characteristics and needs of the historic center in relation to the broader context of the city.
- **Analyze the fabric of the historic center in detail**: Assess its potential and challenges by examining buildings (both occupied and abandoned), state of preservation, housing conditions, existing and potential outdoor spaces, etc.
- **Study the uses of the historic center**: Understand how the historic center functions in relation to the city as a whole.
- **Maintain a functional mix**: Ensure a diverse range of uses within the historic center to avoid specialization and the dominance of a single-use.
- **Develop a Master Plan**: Create a coordinated plan with long-term objectives for the historic center.
- **Create a detailed plan**: Provide specific guidance for the conservation of buildings and outdoor spaces, applicable to both small-scale and large-scale interventions.
- **Direct major investments**: The public administration should guide interventions by large investors to prevent them from concentrating in specific areas. Investments should be spread across prioritized areas and sectors.
- **Avoid focusing on a single area**: Implement improvements simultaneously across various locations within the historic center. Instead of concentrating efforts in the most central area, initiate smaller-scale projects in different areas to act as catalysts for the broader redevelopment of the historic center.

2.4.1 Zoning: Approach to Zonation and Specific Regulations on Historic Buildings

The Urban Plan for the historic center typically requires a different and more detailed approach than a Master Plan for new parts of the city. It necessitates specific guidelines for each building and urban space based on thorough pre-existing studies.

The studies outlined in Chap. 6 aim to demonstrate the types of investigations needed to develop the Plan's prescriptions at the detailed level required for the historic center.

In principle, the Plan could be structured as follows, investigating the **Consistency of the Historic Center** with a series of maps in order to collect:

- Construction phases, buildings, and outdoor spaces.
- Networks, underground utilities, and lighting systems for outdoor spaces.
- Mobility and pedestrian flow.
- Proper@ @ty regime.

- Building materials, paving of outdoor spaces, and elements of urban furniture.
- Characteristic elements of buildings and urban identity.
- Use of public and private buildings and outdoor spaces (ground floor, upper floors).
- Services, trade, and activities.
- Housing condition analysis.
- Synthetic conservation status of buildings and public and private outdoor spaces.
- Conservation status of the building facades.

2.4.2 Plan Documents

Based on the information gathered and the analysis conducted, the Plan must define its strategic objectives (Figs. 2.1 and 2.2). Tentatively, these objectives may include:

Enhancing accessibility: Improve accessibility to the historic center by promoting the use of soft modes of transport, such as cycle paths, footpaths, and hiking trails.

Maintaining perspective views: Preserve and maintain visual connections between the historic center and the surrounding landscape, avoiding new buildings that could obstruct these views and controlling the height and dimensions of any new structures.

Preserving historic fabric: safeguard the historic fabric and the relationships between outdoor spaces and buildings, including the compactness of volumes, the interaction between streets and buildings, access stairs, fixed furnishings, walled gardens, vegetable gardens, and boundary walls.

Prioritizing Degraded Areas: Focus on interventions in the most degraded or at-risk parts of the historic fabric, including both buildings and outdoor spaces.

Maintaining Architectural Consistency: Preserve the architectural and material integrity of existing buildings.

Respecting Building Rhythm: Ensure that building elevations are in harmony with the surrounding public spaces.

Considering Compatible Uses: Explore changes in building uses that align with their characteristics, prioritizing social housing, temporary residences, and light craft activities to encourage the repopulation of the historic center.

Enhancing Public Spaces: Revitalize key public areas, such as squares, and consider architectural projects that repurpose historic building ruins for public benefit.

Regulating Parking: Implement regulations for parking, including the creation of loading and unloading zones within the historic center.

Preserving Historic Pavements: Maintain the historic paving of public outdoor spaces, ensuring consistency with the verification of utility networks and subsurface services.

Improving Public Lighting: Enhance public lighting throughout the historic center to ensure better visibility and safety while respecting the area's historical character.

Considering Technological Innovations: Explore potential technological upgrades, such as installing cabling within the historic center, while ensuring that these enhancements do not compromise the area's historic character.

Based on the gathered information and analysis, the Plan will need to define its strategic objectives. Tentatively, these may include:

Enhancing Accessibility: Improve access to the historic center by promoting the use of soft modes of transport, such as cycle paths, footpaths, and hiking trails.
Maintaining Visual Integrity: Preserve the visual perspectives between the historic center and its surrounding landscape. Avoid new buildings that obstruct these views and regulate their height and dimensions.
Preserving Historic Fabric: Protect the historic fabric and the relationships between outdoor spaces and buildings. This includes maintaining the compactness of volumes, the interaction between streets and buildings, access stairs, fixed furnishings, walled gardens, vegetable gardens, and boundary walls.
Prioritizing Degraded Areas: Focus on interventions in the most degraded or at-risk parts of the historic fabric, including both buildings and outdoor spaces.
Preserving Architectural Consistency: Maintain the current architectural and material integrity of existing buildings.
Respecting Building Rhythm: Ensure that building elevations harmonize with the surrounding public spaces.
Considering Compatible Uses: Explore potential changes in building uses that align with their characteristics, prioritizing social housing, temporary residences, and light craft activities to encourage the repopulation of the historic center.
Enhancing Public Spaces: Improve key areas, such as squares, and consider architectural projects that repurpose historic building ruins for public benefit.
Regulating Parking: Implement parking regulations, including the creation of loading and unloading zones within the historic center.
Preserving Historic Pavements: Maintain the historic paving of public outdoor spaces, ensuring compatibility with utility networks and subsurface services.
Improving Public Lighting: Enhance public lighting throughout the historic center.
Considering Technological Innovations: Explore potential technological upgrades, such as installing cabling within the historic center, while respecting its historic character.

The Plan documents may consist of the Plan Tables and the Technical Implementation Regulations.

The Plan Tables contain comprehensive information related to the strategies implemented, zoning for the area, building uses, and categories of interventions for individual constructions, based on the detailed analysis previously described.

For example, the Plan Tables might include the following:

General Plan Strategies: Outline the specific strategies to be implemented based on the information collected and analyzed during the Preliminary Studies and survey phase.

Identification of Intervention Units: Define the minimum units of analysis and the types of interventions required. For certain types of interventions (e.g., seismic upgrades, reconstruction of multiple adjacent buildings), the project units may consolidate several Architectural Units.

Identification of Interventions for Outdoor Spaces: Specify the interventions needed for public outdoor spaces.

Designation of Building Interventions: Provide detailed indications for the interventions to be performed on buildings.

The Technical Implementation Rules, presented in a written report, could include the following:

Strategies and Plan Addresses: Outline strategic directives, the theoretical framework of the Plan, and general criteria for interventions in the built environment.

General Provisions: Provide definitions, Plan outlines, and other foundational elements.

Implementation Provisions: Detail areas of application for the Plan, zoning related to the general strategies, categories of interventions for the built environment, use regulations, minimum quality standards for project submissions, and documentation requirements for completed works.

Specific Provisions: Address interventions related to outdoor spaces, materials, and building elements.

2.4.3 Checklist for the Evaluation and Approval of the Projects by the Local Authorities

In general, the term "checklist" refers to a comprehensive list of tasks to be completed or items to be verified during a specific activity. This document can streamline the workflow and also serve as a retrospective verification tool. Ensuring that all items on the list are checked off is the most reliable method for completing a task that involves multiple steps.

A specific checklist that outlines the entire process can assist administrators in verifying whether submitted projects meet documentation requirements and professionals and residents in ensuring compliance with all requests.

The checklist can be organized into two levels: quantitative and qualitative. The quantitative level focuses on verifying the presence of all required documents, while the qualitative level assesses whether preservation issues and other criteria outlined in the Plan are met.

To create an effective checklist, it is essential to first identify the objectives against which the evaluation will be conducted.

Different checklists can be designed for various users with distinct goals. Administrators, for example, can use them to assess whether the Plan is progressing as intended and to review submitted projects.

For instance, key objectives for Kumayri Reserve from the administrators' perspective might include:

Preserving Urban Resources: Identifying critical issues and opportunities within the urban system.
Encouraging Public Engagement: Promoting the involvement of the local population in the planning and implementation processes.
Leveraging Local Heritage: Utilizing the local heritage to drive sustainable development.

Management can be supported by a self-assessment grid aligned with these macro objectives, detailing individual actions and criteria. Each step to be verified is framed as a question, and questions can be assigned scores to reflect the level of satisfaction with specific criteria (e.g., on a scale of 1 to 3).

To ensure a balanced evaluation, the process can also involve a group of external stakeholders through peer review.

During the implementation of the Masterplan, the abovementioned indications can be a useful tool for checking the completeness of documents from a quantitative perspective. From a qualitative perspective, compliance with the suggested criteria can be verified according to the recommendations contained in the following chapters. To determine whether submitted projects meet the requirements, a list can assist in the verification process. A specific checklist for professionals helps ensure the submission of uniform material for evaluation.

References

1. Colette A (2007) Climate change and world heritage. Report on predicting and managing the impacts of climate change on World Heritage and Strategy to assist States Parties to implement appropriate management responses. UNESCO World Heritage Centre. https://unesdoc.unesco.org/ark:/48223/pf0000160019.locale=en. Accessed 15 May 2024
2. Fiorani D (2019) Il futuro dei centri storici. Digitalizzazione e strategia conservativa. Edizioni Quasar, Roma
3. ICOMOS (2019) Climate change and cultural heritage working group. The future of our pasts: engaging cultural heritage in climate actions, July 1. ICOMOS, Paris. https://openarchive.icomos.org/id/eprint/2459/1/CCHWG_final_print.pdf. Accessed 14 April 2024
4. UNESCO (2008) Policy document on the impacts of climate change on world heritage properties. https://whc.unesco.org/uploads/activities/documents/activity-397-2.pdf. Accessed 12 May 2024
5. UNESCO (2011) Recommendation on the historic urban landscape. https://whc.unesco.org/uploads/activities/documents/activity-638-98.pdf. Accessed 12 Aug 2024
6. UNESCO (2015) The world heritage convention. Policy for the integration of a sustainable development perspective into the processes of the world heritage convention. https://whc.unesco.org/en/compendium/55. Accessed 12 April 2024
7. UNESCO (2023) World Heritage Centre. Policy document on climate action for world heritage (11/2023). https://whc.unesco.org/en/documents/204421. Accessed 14 May 2024
8. United Nations (1987) Report of the world commission on environment and development: our common future. Oxford University Press, Oxford

9. United Nations (2015) Transforming our world. The 2030 Agenda for sustainable development. https://sustainabledevelopment.un.org/content/documents/21252030%20Agenda%20for%20Sustainable%20Development%20web.pdf. Accessed 21 May 2024

Chapter 3
The Matter of Historic Centers in Armenia

Abstract This chapter examines the situation of "historic centers" in Armenia through three case studies: Gyumri, Ashtarak, and Goris. Each city, distinct in its geographical location, landscape context, size, history, and characteristics, represents the diverse challenges and conditions encountered by "historic centers" amidst a rapidly changing environment. The chapter not only provides a historical overview and current analysis of these three sites but also proposes strategic actions aimed at fostering their development while preserving their unique historical and architectural attributes. Additionally, it explores the tools available for the protection of historic cities, emphasizing the benefits of a heritage-conscious approach. These benefits extend beyond the preservation of historical memory, contributing significantly to the overall revitalization and sustainable growth of the city.

Keywords Gyumri · Ashtarak · Goris · Guidelines for Armenian historical centers · Instruments for historical center protection

3.1 Introduction

International experience demonstrates a clear correlation between revitalization projects targeting historic city cores and underutilized land and a city's capacity to attract talent and business investment. Numerous cities in both developed and developing nations have effectively leveraged their historic centers and underutilized spaces to become talent hubs, drawing global leaders in the knowledge industry and securing foreign direct investment while also fostering vibrant local business development.

Cities preserve people's memories and traditions within their historic cores, reinforcing a unique local identity amid globalization. Preserving architectural and urban heritage serves as a way to honor the past and as a sustainable practice for the present and future. The restoration of ancient buildings minimizes land consumption, and

M. Giambruno et al., *Heritage-Based Tools for the Sustainable Urban Development of Historic Centers*, PoliMI SpringerBriefs, https://doi.org/10.1007/978-3-031-75259-9_3

reducing the use of new construction materials helps control air pollution. Additionally, this approach creates new job opportunities by employing skilled workers in local building traditions.

The definition of a historic city in international charters is a concept that encompasses urban forms from the past that have persisted into the present, carrying various meanings at different spatial scales. It includes elements arising from the intrinsic characteristics of the built environment alongside the values, attitudes, planning, and management designations associated with it. This definition also emphasizes the significant presence of the past, interpreting "historical" as encompassing all cultural, architectural, and urban expressions that are acknowledged as relevant and that reflect the social and cultural life of a community.

The management of historic cities requires balancing conflicting values and evolving interpretations of architectural heritage while developing tailored management policies. In Armenia, historic cities are characterized by mixed land use, a hierarchy of streets and open spaces, and a reliance on passive systems for thermal comfort. These cities are notable for their resilience to change, traditional architectural styles, and harmonious integration with the surrounding environment.

Almost all cities were developed on the foundations of ancient historical settlements. Even newly established urban areas, which are exceptions, typically have a designated "old" city center, around which new districts have gradually formed, often characterized by buildings of lower value. To prevent the historical parts from being overshadowed by the continuously expanding urban structure and to ensure organic integration with future development, it is essential to first study and assess the urban environment. This involves identifying and distinguishing historical layers and areas based on their respective degrees of value.

Next, it is crucial to develop different approaches for the use of historically valuable cores—ranging from removal, preservation, conservation, and renewal to further development (revitalization). Throughout these approaches, the primary considerations should be preserving the integrity of the environment and adhering to the principle of optimal perception as primary and decisive during all the listed uses.

To achieve the ultimate goal of restoring the city's historical environment, it is recommended to create programs with varying priorities, including short-term, medium-term, and long-term plans.

3.2 Between Abandonment and Gentrification. Three Historic Centers as Examples

A study of historic settlements in Armenia reveals a diverse range of conditions, from well-managed and protected areas to those that are nearly abandoned and have lost much of their value. To illustrate these differences, we consider three cities located in various regions of Armenia, each with distinct natural, landscape, and cultural contexts, as well as differing in size, functions, and characteristics: Gyumri, Ashtarak, and Goris.

3.2.1 Gyumri

Gyumri is the second-largest city in Armenia today. Located in the northwest of the country, on the left bank of the Akhuryan River, it is situated 126 km from Yerevan. Historically known as Kumayri, Gyumri has been inhabited since ancient times.

The Armenian historian Ghevond is among the earliest chroniclers to mention Kumayri [1, p. 389]. Archaeological excavations reveal that a settlement existed in the area as early as the 3rd millennium BC, and it has remained continuously inhabited to this day.

The earliest bibliographic reference to Kumayri dates back to the eighth century, describing it as a fortified residence of a duke. During the Bagratuni dynasty, Kumayri developed into a strategic fortress city where crafts and trade flourished.

Known by various names, including Kumayri, Kumiri, Gyumiri, and finally Gyumri, the city suffered significant destruction from the Tatar-Mongol invasions and endured further devastation during the Turkish-Persian rule. It experienced a revival only after Eastern Armenia was reunited with Russia.

In 1837, the first master plan for the settlement was developed, leading to the creation of the new city, renamed Alexandrapol, which expanded east of the old Kumayri. Despite the challenging terrain, Alexandrapol was designed with a regular grid pattern, featuring streets running north–south and east–west, dividing the city into large (100 m × 150 m) and small (50 m × 75 m) quarters. The buildings were primarily 1- to 2-story structures made from black and red tufa stone.

The city plans from 1841 and 1845 continued to adhere to these original principles. The central public space, the Market Square (renamed Vardanants Square in 2013), was surrounded by shops, workshops, administrative buildings, and churches [4]. In 1840, Alexandrapol was designated a provincial city, and by 1850, it had gained the status of a city of provincial significance within the Yerevan province. Due to its advantageous geographical location at the intersection of international trade routes, Alexandrapol emerged as a prominent commercial and industrial hub in Transcaucasia.

Today, Gyumri is the regional center of Shirak, known for its rich historical, architectural, ethnographic, and cultural heritage. The city is situated in a cold climatic zone, with average temperatures ranging from − 5 to − 12 °C in January and around 16 °C in July. The area's land is conducive to landscaping, and there are numerous local stone mines, particularly for tuff. Many buildings in Gyumri are adorned with sculptures and feature windows and entrances framed by bow-shaped arches. The use of black and red tuff gives the city its distinctive architectural character.

Residential Architecture

The residential houses of Alexandrapol were either village-type or city-type. Village-type houses, built in the first half of the nineteenth century, largely followed the composition of traditional Shirak folk dwellings. These were "complex" houses, with a main structure featuring a "thousand" roof, grouped with auxiliary rooms around an internal covered courtyard, forming a single, unified volume. In the second half of the nineteenth century, 1- to 2-story city-type residential buildings were constructed, categorized into single-row or double-row room arrangements. These black terracotta tufa houses had patios on the courtyard side and elegant arched gates facing the street. Their facades were notable for their varied designs, achieved through decorative red tuff elements and emphasized structural edges [5].

The architectural features of the buildings also reflected the social status of their owners. Wealthy residences were characterized by garnet eaves, carved stone gutters, window screens, and luxurious details such as wooden latticework on balconies. Preserving the historical center and its buildings could help transform Gyumri into a modern, vibrant, and sustainable city, in alignment with the 2030 Agenda, addressing the decay and underutilization that have affected the city's oldest parts since the devastating earthquake (Figs. 3.1, 3.2, 3.3 and 3.4).

Strategies for Preservation

To achieve this goal, appropriate tools must be employed to raise awareness of the historic center, its potential, and its current challenges. Clear strategies should be defined for the city, particularly for the historic center, guiding transformations, including the introduction of new functions, services, or buildings. It is also essential to provide direction on proper intervention methods for both large investors and small property owners.

Zoning guidelines in the Master Plan alone are insufficient to preserve a historic center. Specific prescriptions, derived from detailed studies of the ancient core and its built environment, are crucial for ensuring the conservation of the historic center while adapting to contemporary living needs [11, pp. 17–63].

Fig. 3.1 Gyumri historic center. In the background, some buildings remain in a state of ruin. In the foreground, a public space has been created from the demolition of buildings destroyed by the earthquake (*Photo* S. Pistidda, 2022)

Fig. 3.2 Gyumri historic center. In the areas furthest from the city's main square, many roofs are still made from temporary materials (*Photo* S. Pistidda, 2022)

Fig. 3.3 Gyumri historic center. Notice the extensions, replaced fixtures, and added elements (*Photo* S. Pistidda, 2022)

Developing preliminary studies is essential for creating a detailed plan for the historic center. These studies should range from basic cartography to graphic scales and study tables, identifying the potential and challenges of the ancient core.

The goal is to define the main issues facing the historic center of Gyumri, establish priority interventions, and outline strategic functions for revitalizing Kumayri. Furthermore, specific legislation on urban decorum and a management plan should be developed to ensure lasting effects. It is also crucial to study best practices in building conservation and provide practical suggestions to residents for simple maintenance work on their properties.

Fig. 3.4 Gyumri historic center. Notice the one-story houses and the added stairs to overcome the different levels of the road surface (*Photo* S. Pistidda, 2022)

Protected Status and Zoning

The central core of the city, known as the "Kumayri" historical-architectural reserve, is designated as a specially protected area. This area encompasses lands of aesthetic, ecological, scientific, and historical-cultural significance and is subject to specific legal regimes for land use. These include gradual and complex design support zones, as well as urban planning, historical-cultural, and engineering-geological restrictions [10, pp. 23–256].

Several zones have been defined, covering the entire area:

- Regulatory Zone for the Preservation and Construction of Historical and Architectural Monuments.
- Potential Excavation Zone.
- Construction Restriction Zone.
- Natural Environment Restoration Zone.
- Road Communication Zone.
- Engineering Infrastructure Zone [6].

To effectively regulate these functions, it was necessary to generate proposals for their adaptive reuse. A methodology must be developed to implement measures aimed at protecting historical heritage from loss or alteration during the process.

Simultaneously, an action plan has been proposed to translate these theoretical provisions into practical applications, taking into account the needs of all stakeholders involved in each decision-making process and establishing a management board that includes administrative members, investors, specialists, and the local population.

3.2.2 Ashtarak

Ashtarak boasts a history spanning thousands of years, with evidence including Bronze Age statues, settlements, tombs, and other monuments. Mentioned in Armenian sources since the ninth century, the city was part of the Ayrarat and Aragatsotn regions of Greater Armenia [2].

Between the eleventh and fifteenth centuries, the Aragatsotn region, along with other historical Armenian territories, faced invasions by the Seljuks, Mongols, Ak Koyunlu, and Kara Koyunlu. From 1930 to 1995, Ashtarak was the center of the Ashtarak region, and in 1963, it was granted city status.

While Ashtarak contains protected districts and individual monuments, it lacks a designated preservation area like the Kumayri reserve in Gyumri.

The city is particularly notable for its religious structures, such as churches, and sections of streets in the historical core, which are lined with significant residential buildings. Various institutions of unique value have been relatively well-preserved in this area. The historical part of the city includes houses where prominent figures of Armenian history and culture were born and lived. These monuments, along with the urban development and the picturesque nature of the Kasakh Valley, enhance the city's historical significance and contribute to its unique historical character and architectural identity. The construction of Ashtarak in that era was based on three main elements: the residential house, the adjacent plot of land, and the street, lane, or dead end [12].

Among the historical monuments of Ashtarak, the most notable are the churches of Tsiranavor, Karmravor, and Marine. Tsiranavor ("Orange"), dating back to the fifth century, is situated on the rocky bank of the Kasakh River. Although it is currently in a dilapidated state, it is safeguarded from further destruction by stone supports on three sides.

Many ancient buildings have been preserved in the city. Since the end of the eighteenth century, low village structures have been replaced by 1- to 2-story buildings. The most notable concentration of historic residential buildings is found on Ghapantsyan Street in the city's central core. Although these traditional buildings are somewhat hidden from view by high-rise structures, a row of local buildings with charming wooden balconies remains on both sides, alongside newer constructions. Some of these historic buildings are in poor technical condition [13] (Figs. 3.5, 3.6, 3.7 and 3.8).

Cities with original and unique characters gain significant aesthetic value, providing an attractive living environment and appealing to visitors. Adapting changes to the historic environment while conserving heritage requires a flexible approach to limit alterations to historic resources.

A well-preserved historic environment can stimulate tourism, supporting the local economy, creating jobs, and serving as an educational resource about local history, construction methods, and traditional lifestyles.

Architectural designs and proposals for the reconstruction, rehabilitation, and modernization of heritage structures, when considered holistically, have shown

Fig. 3.5 The poster for a workshop held in Ashtarak in 2017 for the conservation of the historic center, organized between NUACA and POLIMI. Professors M. Boriani, M. Giambruno, N. Petrosyan, M. Bunatyan (© M. Giambruno, 2017)

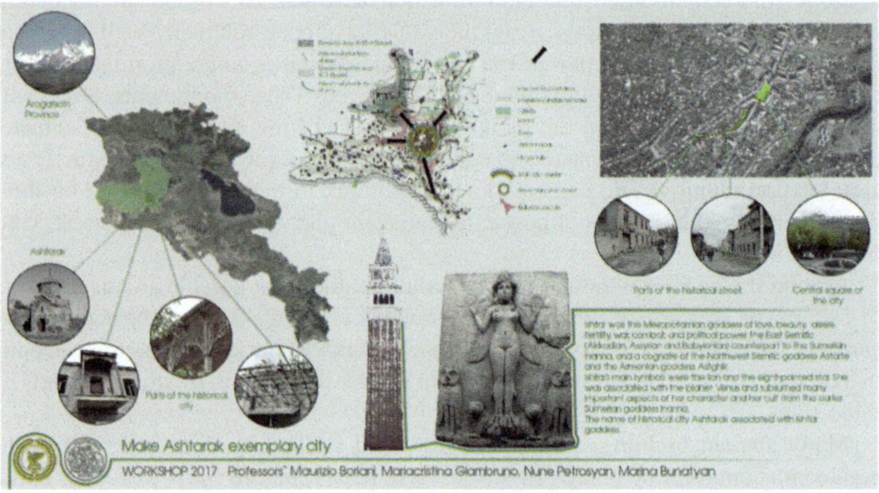

Fig. 3.6 Identification of the distinguishing elements. Students project. Workshop, Ashtarak, 2017. Professors M. Boriani, M. Giambruno, N. Petrosyan, M. Bunatyan (© M. Giambruno, 2017)

significant results internationally. This approach includes the buildings themselves and their surrounding historical landscapes, elements of daily life, agricultural activities, crafts, and engineering structures. A city with a unique architectural style offers valuable potential for enhancing its status and promoting its prosperity.

In preserving the identity and historical value of the environment, flexibility is crucial to maintain the unity, continuity, and appropriate combinations of style, mass, and scale within the historical context. This approach involves implementing moderate interventions that respect material heritage and existing structures. Preserving and reusing these elements within their original context is more valuable than recreating them from scratch.

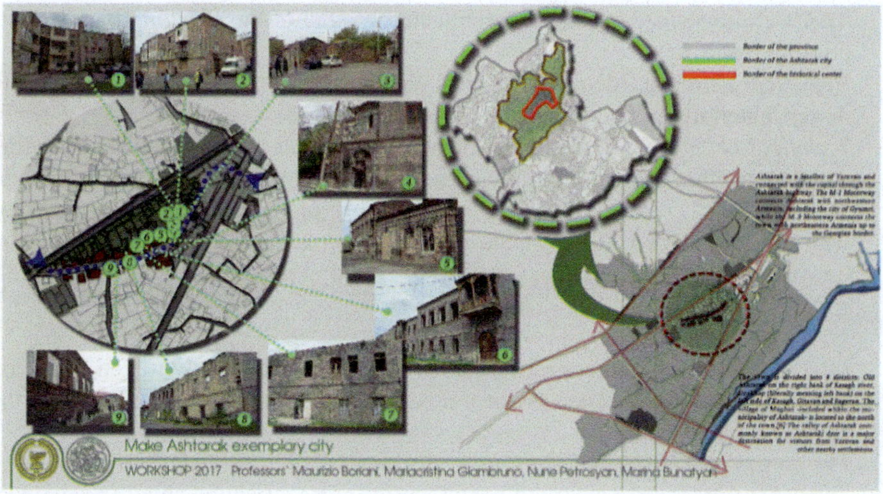

Fig. 3.7 Mapping of Ashtarak historic center. Students project. Workshop, Ashtarak, 2017. Professors M. Boriani, M. Giambruno, N. Petrosyan, M. Bunatyan (© M. Giambruno, 2017)

Fig. 3.8 The historic center of Ashtarak (*Photo* M. Giambruno, 2017)

A city, as a living organism, evolves over time, blending historic and new elements. Consistency in preserving the original appearance of materials, structural systems, architectural details, and ornaments is essential. This includes maintaining the general mass and shape of buildings and their environment, which helps identify them as characteristic of a particular period.

Preservation is a sustainable approach that promotes the reuse of historic buildings and existing urban environments. Incorporating energy-efficient solutions in modernization can reduce utility costs, extend material lifespan, and mitigate negative environmental impacts.

Before determining the approach—whether preservation, restoration, repair, or reconstruction—a comprehensive evaluation process should be undertaken based on the following criteria:

Architectural value: Assess the architectural value of the structure based on its original features.
Historical Features: Identify historical elements that may be concealed beneath newer materials applied to parts of the structure.
Technical Condition: Evaluate the physical condition of the structure, including any damage that could compromise its structural integrity.
Material Replacement: Consider the possibility of replacing original materials with similar ones or using new materials if the original ones are no longer available.

As a guideline, the following set of international rehabilitation standards is proposed:

Intervention Areas/Structures: Areas or structures should be utilized based on their historical significance or for a new function, ensuring that changes to the environment are minimal.
Preservation of Historical Materials: The removal of historical materials or alteration of features and spaces within intervention areas/structures should be avoided. Key characteristics, exterior layers, construction methods, and craftsmanship must be preserved.
Authenticity of Intervention Areas/Structures: Each intervention area/structure should be regarded as a physical record of its time, location, and functional significance. Modifications should not create a misleading impression of historical development by adding assumed features or architectural elements from other buildings.
Repair rather than Replacement: Historic elements in poor technical condition should be repaired rather than replaced. If replacement is necessary, the new parts should match the original in form, color, texture, and, where possible, materials and other characteristics. Replacement should be guided by preserved documentary materials and photographs.
Preservation of Historical Changes: Over time, many intervention areas/structures undergo changes. These changes, which hold historical value, should be preserved as additions that reflect the period during which they were made.
Treatment of Historic Materials: Chemical or physical treatments that could damage historic materials (e.g., sandblasting) should be avoided. When cleaning surfaces, use the mildest methods possible, such as washing.
Distinguishable New Additions: New additions must be clearly distinguishable from the old while being compatible in terms of mass, size, scale, and architectural features.
Future removal Considerations: New additions and adjacent or related new construction should be designed in a way that preserves the integrity of the historic setting or structure, allowing for the possibility of future removal without damage.

3.2.3 The Old Town of Goris

Currently, with the onset of uncontrolled urbanization processes, the issues of modernizing urban infrastructure, adapting it to contemporary needs, and enhancing operational efficiency have become pressing worldwide. In this context, the concept of smart cities is gaining traction, gradually expanding from major metropolises to medium and smaller cities. This concept offers pathways for both preserving historical and cultural heritage and achieving sustainable territorial and infrastructural development.

In the definitions of sustainable urban development, terms such as sustainable, green, smart, resilient, adaptable, and comfortable are considered equally important. While these terms describe different characteristics, they share a common goal: to create a favorable, accessible, ecologically safe, secure, and attractive living environment. This environment should meet the fundamental needs of residents—such as housing, work, and recreation—while preserving cultural heritage accumulated over centuries for the benefit of future generations.

Environmental sustainability is a crucial component of urban development, especially in cities reliant on tourism and natural resources. Given the scarcity of natural resources, it is essential to ensure their safe and renewable use. Balancing development with conservation is fundamental to achieving sustainable urban growth.

Goris is first mentioned during the Urartian period. King Rusa I, who reigned in the eighth century B.C., left a cuneiform inscription in which he listed 23 countries he had conquered, including one called "Goristsa". Scholars believe that this "Goristsa" is the same as present-day Goris [3].

During the Middle Ages, the town-settlement of Goris was located in the eastern part of present-day Goris, on the left bank of the Goris River. This area was known as Old Goris and corresponds to one of the villages named Goru and Goraik mentioned by Stepanos Orbelyan in the thirteenth century. In medieval times, Goris served as a crucial junction on the busy trade route from Nakhijevan to Agulis and Partav, passing through Zangezur and Syunik to the Near-Caspian steppes. This route was vital for transporting salt from Nakhijevan.

In this context, the comprehensive study has led to the following recommendations for Goris, one of the most beautiful and distinctive cities in Armenia. As the southern gateway to the republic and a key hub of the Syunik marz, Goris has been recognized for its significant role, position, and potential. It has been identified as a primary cultural center of both regional and national importance.

Situated at the crossroads of key historical routes, the city has experienced rapid growth, evolving from a small rural settlement into a vibrant commercial, educational, and cultural hub. It boasts its own traditions and a unique architectural identity, characterized by distinct sections from three different periods. These sections are clearly delineated yet seamlessly connected by transitional layers [3].

Today, despite retaining its historic values and a wealth of cultural heritage monuments—ranging from worship and defense structures to administrative, private, and public buildings—the city has undergone significant transformation due to changes in the geopolitical landscape. Once a pivotal hub at the intersection of major routes such as the Yerevan-Artsakh and Yerevan-Meghri-Iran highways, located in the Goris (Vararakn) River Valley and 235 km from the capital, Yerevan, the city has now become a more isolated location with limited transportation access.

The study of the city's urban morphology reveals that the most valuable part of Goris is the Old City and the historical core of the current town center. This area features distinctive architecture, including unique one- and two-story houses with masonry facades adorned with a special plastering technique known as "barqash". Stone fences and arched gates, which display arches with various centers and radii, are harmoniously integrated with both the natural and urban landscapes surrounding them.

The stone pyramids of Old Kores, located to the east of the modern town, are among the most captivating sites in Goris. This cave settlement features defensive caves that were accessible only by ropes. Some multi-story cave houses were connected by internal corridors, with upper floors typically featuring wooden balconies extending over the cliff. Over time, the caves were expanded and modernized with arched, curved facades and various amenities, such as toasters, ovens, and cupboards. From the nineteenth century onward, stone houses began to be constructed in Goris, and the caves, known locally as "Krataker", were repurposed as additional living spaces. In "flat" Goris, most houses are two stories high. The lower floor typically served as a kitchen, storage area, or for other practical purposes, while the upper floor, where the residents lived, consisted of a porch and either one large or two smaller rooms. These houses were distinguished by their arched porches and elegant wooden balconies. They were constructed from high-quality basalt, with roofs covered in wooden tiles.

The old town is home to the 4th-century basilica of Surb Hripsimé, which was renovated in the sixteenth century. As mentioned earlier, a significant portion of Goris's city core is designated as a historic district reservation, with many single-family homes recognized by the Ministry of Culture as national heritage and subject to strict preservation guidelines. From an architectural and urban planning perspective, the most remarkable features of Goris include its distinct architectural structures, the streets they form, the historical center, natural monuments such as grottos, and the combined areas of Old and New Goris. According to the historic master plan, a rectangular road network with streets 16–21 m wide divides the area into nearly equal rectangles measuring 106 by 212 m.

The condition of this neighborhood is poor due to a lack of preservation, as many historic buildings, despite existing regulations, have been modified beyond recognition and have lost their historical value. Even the restrictions on building height have been ignored. Many houses have added extensions to their facades, completely altering the historic appearance of the buildings. Additions to properties have also led to the disappearance of traditional property setbacks that were commonly used as gardens. Numerous historic properties have erected stone enclosures that obstruct the view of the buildings. Balconies, once a characteristic feature, have often been removed or converted into additional rooms [7, p. 38].

In the 1980s, mid-rise concrete residential buildings were constructed along the boundary of the historic district. These developments occupy large superblock lots that were once integral to the historic fabric of the city. Many of these buildings now suffer from outdated and dilapidated facilities, communal spaces, and infrastructure.

Preservation of the historic district would involve the restoration of historic facades, preservation of the original lot typology and connections, and overall enhancement of the district's historic qualities that existed during its formative period. This approach aims to create a city center that embodies a sense of destination, arrival, and placemaking. The city center would serve as a gateway to Goris, around which a hub of diverse public, private, and recreational activities could flourish.

Currently, this area lacks identity and a sense of place and is generally unappealing. An effective way to protect the gradually diminishing features of the historic core is to recognize the importance of developing the area as a gateway to Goris, using urban design elements such as materials, finishes, building proportions, and scale adjustments that reflect the historic context and heritage. For example, introducing a change in the pavement before entering the city center can signal to drivers that they have entered a distinct area, encouraging them to slow down.

A human-scale urban environment can be created using elements such as streetlights, benches, trees, and ground-floor commercial spaces, which will provide the area with a much-needed identity and invite visitors to stop and enjoy the surroundings. As part of a holistic and inclusive approach to master planning, historic buildings play a crucial role in preserving the identity of Goris' historic core. Most of the properties identified as having historic value are already on the Ministry of Culture's Historic Preservation list, which includes specific control parameters established to protect and preserve these buildings. However, some properties have been so altered that they have lost not only their character and beauty but also their historical significance.

The best approach for preserving and adaptively reusing historic properties involves transforming neighborhoods into vibrant, lively centers with tastefully preserved buildings and landscaping. This is not merely a gentrification of the historic environment—often resulting in the erasure of historic layers and the application of incongruent new features—but rather, it aims to achieve the vision through the following steps:

Restoring the architectural integrity of defaced historic properties: This includes façade restoration, plastering, and painting; replacing windows; installing clay roof tiles; restoring architectural details; and repairing or rebuilding wooden staircases, railings, and columns.

Protecting and preserving architectural character and integrity of properties that are considered of historic value.

Removing solid fences or other obstacles: This will enhance the visibility of historic buildings and create a safer, more engaging streetscape for pedestrians.

Narrowing streets to reduce vehicular speed and *widening sidewalks* to encourage pedestrian activity.

Introducing stone paving on streets, stairs, and sidewalks to enhance the historic atmosphere.
Beautification and landscaping to infuse seasonal vibrancy with colorful greenery.
Proper street lighting, using light posts and wall sconces that complement the scale and historic character of the street, enhancing both the evening ambiance and safety.
Encouraging local entrepreneurship by promoting small and medium-sized enterprises (SMEs) such as gift shops, restaurants, bakeries, bed-and-breakfasts (B&Bs), and other local artisan businesses. This initiative aims to attract visitors while providing income opportunities for residents.
Improving infrastructure, including gas, water, sewer, electricity, telecommunications, and drainage channels [8, pp. 27–213].

The importance of cultural heritage in shaping identity underscores the necessity of rebuilding destroyed environments and recovering lost memories. This has significant implications for modern building practices, as it emphasizes the need to respect and incorporate historical context while creating new structures.

Cultural heritage often serves as a distinguishing factor among different communities, yet it is clear that regional architecture has been shaped by a blend of cultural influences, including elements like the Hazarashen roof cover. Preserving cultural heritage through rehabilitation and adaptive reuse is crucial, as it not only sustains the identity of traditional building practices but also facilitates an understanding of their origins and durability in architecture.

The significance of preserving and making use of historic assets becomes particularly apparent when examining historic rural settlements such as Old Khot, Old Shinuhayr, and Old Halidzor around Goris. These mountainous settlements, though abandoned by their inhabitants, have largely retained their urban identity, including their road networks, buildings, and channels. They provide valuable examples of traditional building practices, including the historic roof coverings used for bakery sections of houses. Despite their preservation of architectural and urban characteristics, these settlements face threats from weathering, demolition, and natural hazards. Adaptive reuse, even when it follows a process of gentrification (with all its well-known negative effects and implications), can, on the other hand, trigger the rehabilitation of such heritage, fostering regional economic development and increasing the incomes of the inhabitants [9, pp. 127–221].

Thus, historic preservation goals—such as the protection, enhancement, perpetuation, and use of structures, sites, and districts of historical, architectural, or landscape significance located within Goris City—are a public necessity and are required in the interest of the community's prosperity, sustainability, civic pride, and general welfare. This means that the first step in the gentrification of these historic sites is to designate, preserve, protect, and enhance the structures, sites, and districts that reflect the outstanding cultural, artistic, social, economic, political, architectural, historical, or other heritage of the settlement. Additionally, this approach aims to protect and enhance property values through the stabilization of neighborhoods and public areas, and to increase economic and financial benefits for the settlement and its inhabitants.

These measures are aimed at:

- Promoting the settlement's attractiveness to tourists and visitors.
- Fostering the reuse of abandoned, yet significant historical or architectural structures, sites, and districts (such as the Goris bathhouse).
- Raising the standard of living in the historic district by addressing the community's infrastructural needs.
- Promoting and setting standards for the beautification and restoration of historic districts, which directly enhances their historic value and, consequently, increases the real estate value of these areas.

3.3 Policies and Instruments for Protection

Policies for heritage protection include both international guidelines, such as those from UNESCO and ICOMOS, and local and national legislation, such as the Historic Heritage Protection Law for historic monuments, which define the activities and methods for safeguarding heritage. The instruments for protecting the historical integrity of existing settlements are implemented through actions such as protection, preservation, conservation, restoration, renovation, adaptive reuse, reconstruction, revitalization, rehabilitation, enhancement, and maintenance.

The goal is not to preserve historic buildings in a museum-like state, but to repurpose them in ways that meet the needs of a dynamic city while still valuing its historic identity.

The benefits of these protective measures include:

Identity Protection: The unique vernacular architecture of Goris and neighboring rural settlements—featuring timber frames, red roofs, stylized wooden balconies, and double-arched gates—is a significant resource. If properly preserved, this architectural heritage will enhance and solidify the settlement's status as a center of historical and cultural significance.

Economic Development: Historically well-preserved neighborhoods unique to Armenia can attract more tourism, which in turn supports local businesses and creates jobs. Preservation is a labor-intensive process that relies on specialized techniques, benefiting local workers who possess unique knowledge of traditional building materials. This contrasts with new construction, which often depends on imported materials. Additionally, the lower cost of maintaining existing buildings makes historic properties particularly suitable for small businesses that might otherwise be unable to afford the rent of newer buildings.

Urban Tourism: Visitors can explore historic streets and use their smartphones to navigate the district, identifying architectural styles, significant events, notable visitors, or previous owners. By scanning QR codes on historic buildings, visitors can view historical images that depict how the streets or properties appeared in the past. This interactive experience can boost interest in historic preservation and stimulate economic development throughout the historic district.

Sustainability: Preservation is a sustainable solution as it promotes the reuse of existing historic buildings and the use of the current built environment, rather than new construction that might encroach on undeveloped natural landscapes. It requires fewer imported building materials and often involves energy-efficient retrofits, which can lower utility bills, extend the lifespan of building materials, and reduce water, energy, and carbon impacts. These measures also enhance the return on investment for building owners.

Considering the unique characteristics of Armenian historic centers, initiating the revitalization process will require studying specialized investment management mechanisms. These mechanisms should ensure that they deliver tangible benefits in conserving both the historic city and its inhabitants.

For the Gyumri Architectural Historical Reserve, a special mechanism has been proposed: the formation of a joint-stock company or investment fund. This collective investment mechanism, similar to a Real Estate Investment Trust (REIT), is intended for the restoration and rehabilitation of the Kumayri Historical-Architectural Reserve Museum. Such a model is already established in international practice as an effective tool for managing and funding historic preservation projects.

The purpose of the proposed trust is to attract investments from various investors to develop the territories and generate income (dividends). In this model, shareholders would include not only financial investors but also property owners of historically valuable buildings. These property owners would have the opportunity to participate in the revitalization of the reserve and receive income by joining the joint-stock company with their real estate without needing to sell their properties while continuing to occupy and own them. Additionally, the community, through its property (such as community land), and the state, through its share, would also become members of the company.

References

1. AA VV (1967) Armenian history, Vol. 2. S.e., Yerevan
2. Artsruni G (1890) In the province of Yerevan, Development No. 175
3. Hakobyan T, Bakhshyan M, Barseghyan H (1986) Dictionary of toponyms of Armenia and adjacent regions. Erevani Hamalsarani Hratarakch'ut'yun
4. Harutyunyan VM (1977) The planning and construction of the cities of Eastern Armenia in the XIX century and the beginning of the XX century. PBU, Yerevan
5. Khalpakhchyan O (1971) Civil architecture of Armenia. Strojizdat, Moscow
6. Law of the Republic of Armenia (1998) Protection and use of immovable monuments of history and culture and the historical circle, November 11, No. ZR-261
7. Papukhyan N (1970) The folk architecture of Syunik. Armenian SSR GA Publishing House, Yerevan
8. Petrosyan NG et al (2014) The master plan with zoning plan of Goris city. Armproject OJSC, Yerevan
9. Petrosyan NG et al (2019) Microregional complex urban planning development plan and zoning of Syunik region settlements (Gorisand rural settlements). Armproject OJSC, Yerevan
10. Qalashyan SH et al (2021) Microregional complex urban planning development plan and zoning of Shirak region settlements (Gyumri and rural settlements). Armproject OJSC, Yerevan

11. Qalashyan SH, Aloyan AA, Petrosyan NG et al (2001) The zoning plan of Gyumri historic center. Armproject OJSC, Yerevan
12. Yaralov YS (1947) Ashtarak/Academy of Architecture of the USSR. Institute of History and Theory of Architecture. Under the general editorship of Academician VA Vesnin. Publishing House of the Academy of Architecture of the USSR (Treasures of Architecture of the Peoples of the USSR)
13. Yervand S (1945) The history of Ashtarak. Yerevan (published in 1987)

Chapter 4
Gyumri and Its Historic Core: A "Historical Reserve"

Abstract This chapter presents the results of research conducted on Gyumri, aimed at developing a strategic framework of practices and tools as a foundation for the city's heritage-based revitalization. The chapter begins by reconstructing the historical events that led Gyumri to its current state, followed by an analysis of its current potential and challenges. Gyumri is a planned city, largely preserved in its layout and buildings—a feature that represents its greatest strength. However, it faces many challenges common to "historic centers" in so-called emerging countries: depopulation, deterioration of the oldest parts, lack of job stability among its inhabitants, low income, and inadequate planning policies. Based on the studies, several strategic functions are outlined that could enhance the city's prospects by preserving its historic buildings and making them the focal point for various revitalization processes. Identifying these strategic functions, derived from local conditions and aligned with its real possibilities in a medium-term scenario, can support the proposals of the Historic City Plan. The historic part of the city can fulfill this pivotal role if provided with tools that ensure its preservation while simultaneously allowing for the introduction of new functions, upgrading housing standards, incorporating new buildings, and conserving existing ones. The final part of the chapter discusses some of these strategies, which, specifically designed for Gyumri, could also serve as inspiration in other contexts.

Keywords Plan and tools for the historic center · Potential and challenges of the historic part of Gyumri · Strategic functions

4.1 Gyumri: A Resilient Foundation City

Considered one of the oldest settlement areas in the country [13, 18], the area now known as Gyumri was little more than a village in 1813. Following the Russo-Persian War and the subsequent Treaty of Gulistan, it was annexed by Russia. The construction of a Russian military outpost near the village, intended as a bulwark against the Ottoman Empire, marked the beginning of present-day Gyumri. The

M. Giambruno et al., *Heritage-Based Tools for the Sustainable Urban Development of Historic Centers*, PoliMI SpringerBriefs, https://doi.org/10.1007/978-3-031-75259-9_4

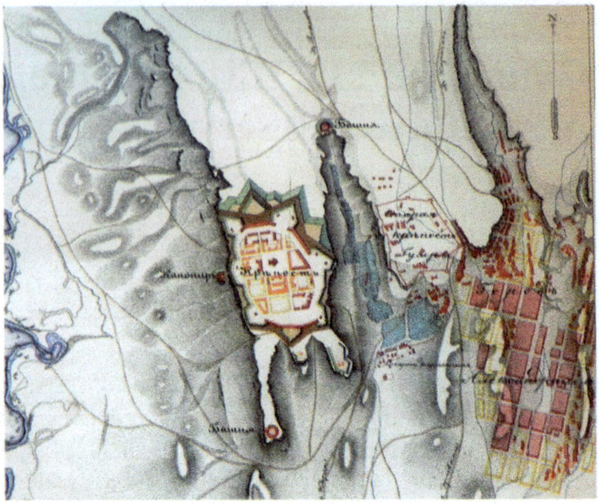

Fig. 4.1 Master plan of Alessandropoli, 1840 (Reproduced from AA [1])

initial nucleus of the city emerged on its outskirts, at the intersection of three major communication routes to the north, west, and south [11, 17].

The layout of the new city is first depicted in a map dating back to 1840 (Fig. 4.1). It is organized in a "checkerboard" pattern, starting with a large central square surrounded by regularly shaped blocks [1].

In the general plan of 1872 (Fig. 4.2), the city expanded according to the principles of the initial urban layout: rectangular lots, nine in the widest part, interspersed with some large open spaces. Near the center, there is a large square where the two main churches face each other. The road network features wide, straight streets that traverse the city from north to south and from east to west [12].

Under Russian rule, Aleksandropol—its name at the time—experienced significant industrial growth, becoming one of the most important cities in the southern Caucasus after Tbilisi and Baku. By 1900 (Fig. 4.3), a new General Plan was needed: the checkerboard layout expanded in all four cardinal directions, while the central areas were redesigned to enhance the city's strength and urban character.

Not even a decade later, Gyumri, located close to the Turkish border, became a refuge for thousands fleeing the Armenian genocide that began in 1915. To accommodate the influx of refugees, the city underwent modifications in its spaces and building uses, with many buildings converted into orphanages and others adapted to meet immediate needs, such as hospitals and schools [7].

The First World War and the Turkish invasion led to a period of economic decline throughout the region. With the establishment of the Republic of Armenia and its return to the Soviet sphere of influence, the city, renamed Leninakan in 1924, experienced a brief period of recovery, which was interrupted in 1926 by a strong earthquake that caused significant damage [8, 16].

In the 1930s, the drafting of a new General Plan led to renewed growth in the city (Fig. 4.4). The city's boundaries expanded northward, and the rigid checkerboard

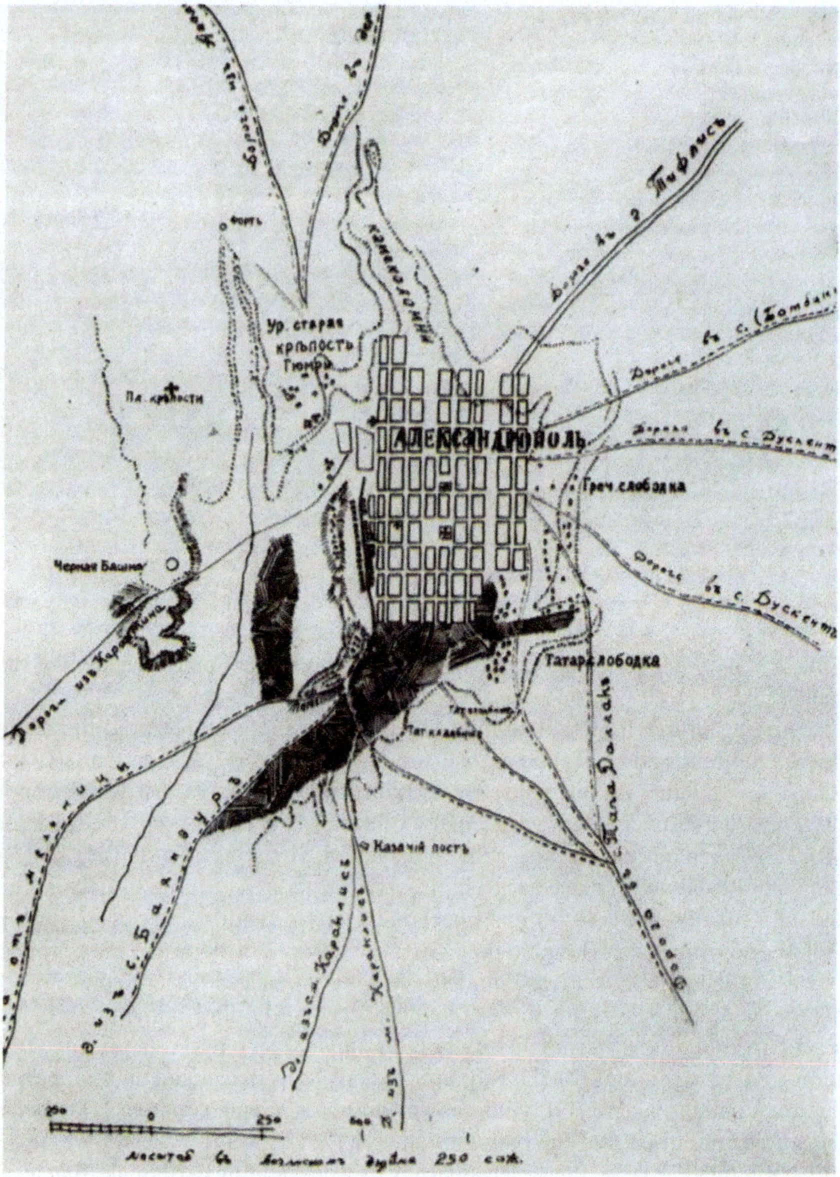

Fig. 4.2 Master plan of Alessandropoli, 1872 (Reproduced from Matevosyan [12])

Fig. 4.3 Master plan of Alessandropoli, 1900 (Reproduced from Matevosyan [12])

layout was modified, even in existing areas, by the introduction of a system of radial streets [5].

Beginning in the 1950s, a significant population increase required the creation of new neighborhoods designed as extensions of the historic city, particularly to the north. These neighborhoods consisted of multi-story buildings that were economically constructed over a few years, many of which did not survive the earthquake of the 1980s. In addition to the new constructions, existing low-density buildings were demolished and replaced with multi-story structures capable of housing more residents. To address the resulting impact on the city's historical heritage, a group of scholars and officials working on the Five-Year Urban Development Plan proposed, as early as the late 1970s, to designate the Kumayri area as a "reserve museum" [3]. This proposal aimed to limit the damage caused by city development. To halt this process, in 1980, the government officially designated Kumayri as a priority conservation area, recognizing it as the center of the city's historical and cultural heritage. The area contains over 1700 buildings spread across approximately 200 hectares, representing about 22% of the entire urban area [2].

On December 7, 1988, the area was struck by a devastating earthquake that leveled the nearby town of Spitak, the quake's epicenter, and severely impacted Leninakan, largely due to the alluvial soil on which it was built. Some estimates suggest that approximately 52% of the buildings were destroyed, particularly those constructed after the 1950s, leading to extensive reconstruction in this part of the city [6]. The historic city also suffered significant damage.

The earthquake caused thousands of deaths and led to the closure of the few industries that, though not destroyed, had become obsolete in terms of production

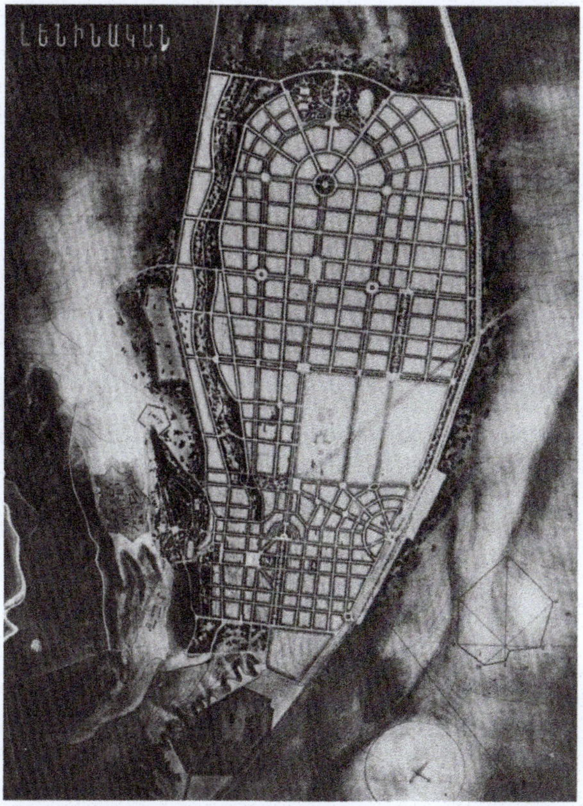

Fig. 4.4 Master plan of Leninakan, 1932–1937 (Reproduced from Dolukhanian [5])

systems. These factors contributed to the city's depopulation, with the population decreasing from 200,000 in 1988 to the current 110,000.

At the time of the Republic of Armenia's independence in 1991, the city had a significantly damaged building heritage and was isolated due to a lack of gas and energy supply.

Damaged by earthquakes and expanded over the years with partial replacements of buildings, Gyumri still retains the urban layout established in the mid-19th century and finalized in the 1930s. A wide boulevard, built on the now-covered riverbed and hosting a linear park, divides the city into two nearly identical zones in size. Some public buildings, including the Museum of Architecture and Urban History, housed in a historic mansion, overlook this boulevard. However, few examples of preserved architecture remain along this main road, as it has been replaced by newer buildings of varying heights, styles, and materials. To the east and west of the central green spine, the areas most clearly reflect the built environment that characterized Alexandropol before and Leninakan later.

The built environment reflects a deeply Armenian identity, expressed through buildings that, unlike the urban layout, are minimally influenced by Russian and

Soviet styles. Local gray, black, orange, and red tuff were skillfully used by local stonemasons, who took advantage of the abundant and easily accessible material in the Gyumri area. The city, crafted by these artisans, showcases construction methods and materials that have provided a continuous thread through its urban development. This distinctiveness has been maintained from its "maximum splendor" in the late nineteenth and early twentieth centuries, through the Soviet era, during which it remained a pivotal center of the country [9].

Except for buildings designed for collective and representative uses, much of the city's heritage consists of red and black tuff cut into square, regular blocks of nearly uniform dimensions. These one or two-story structures have an austere appearance, featuring a sober decorative scheme with cornices, pilasters, and ornamental bands on the street-facing façades.

Today, more than thirty years after the earthquake, the city has partially resumed its life and regained its former role. Particularly around the main square, this "rebirth" is evident through various urban-scale projects, including some internationally recognized initiatives. Notably, the recent pedestrianization of the cross street connecting the main square to the urban park—where the city's main axis concludes—has revitalized the area. The street has become lively once again, with bars, restaurants, and craft shops selling local products.

Conversely, not far south of the main square, the large covered market—once the heart of local commerce—is now a ruin, largely abandoned and unused by local producers. The cost of renting space inside remains prohibitively expensive relative to the meager earnings it generates. Surrounding streets, as well as the ground floors of 19th-century buildings still uninhabitable due to earthquake damage, have evolved into a lively yet impoverished informal market where residents sell whatever they produce, even from their small gardens. In the central areas, it is not uncommon to encounter buildings still in ruins, with collapsed roofs and floors, a consequence of both earthquake damage and prolonged neglect.

These two extremes coexist in Gyumri, reflecting the city's inherent contradictions. Despite this, since the late 1990s, Gyumri has been the focus of numerous reconstruction and revitalization projects, primarily funded by international organizations. These efforts aim to draw residents back to the city and rejuvenate its economy and appeal. The initial projects concentrated on reconstructing multifamily residences in the suburbs, revitalizing infrastructure, and restoring museums and public theaters. They also included the construction of new facilities such as schools, medical centers, and sports complexes. More recent projects, however, have shifted their focus to the city's historical heritage, with an emphasis on promoting tourism, particularly international mass tourism [14, 15].

A wide array of initiatives, often overlapping and sometimes competing, has yet to produce tangible and lasting results in revitalizing the city and encouraging residents to return. Similarly, these efforts have not significantly impacted the conservation of its architectural heritage.

Most interventions have been limited to "exemplary" buildings that, while noteworthy, do not address the broader, systemic needs of the city. What is required is a comprehensive approach that considers the city as a whole and tackles the larger issues at hand. Such an approach could potentially trigger a genuine "rebirth" of Gyumri by solving its underlying problems and effectively enhancing its urban fabric while respecting the city's unique identity and layered history [10].

There could be several reasons for this situation, including a lack of updated discourse on urban restoration, which appears to be underdeveloped both at the governmental level and within Armenian academic institutions. Additionally, insufficient funding, especially in the absence of international aid, has hindered the implementation of planned projects.

What is clear, however, is that efforts to "regenerate" Gyumri largely focus on developing the tourism industry, despite the challenges and limitations this approach presents. Recent initiatives, such as attempts to include Kumayri on the UNESCO World Heritage list—marked by a visit from UNESCO representatives in August 2019—seem to reinforce this trend.

4.2 The Kumayri Historical Reserve: Resources and Criticalities

Gyumri and its historic center represent a unique case in the Armenian context due to their size, the characteristics of the buildings, and the urban layout. These attributes make it a notable example, potentially significant in other geographical contexts as well.

Starting in the 1970s, efforts to protect the historic area of Kumayri led to the imposition of protective restrictions, effectively designating a substantial part of the city as a "museum-reserve". From the 1970s until the 1988 earthquake, a large team of experts—about 50 people—was responsible for preserving the historic buildings, ensuring the conservation of the city's historic core, which had expanded beyond the original city center.

The 1988 earthquake, which caused extensive death and destruction and led to the dissolution of the team responsible for safeguarding the historic center, resulted in numerous problems that contributed to the gradual degradation of its historic heritage. Key issues included the city's depopulation, a decline in economic activities, and a loss of its role as a regional hub.

The earthquake severely impacted the city's outskirts with partial and total collapses of new multi-story reinforced concrete buildings and also caused significant damage to the historic center. Two reports on the condition of the historical heritage immediately after the earthquake revealed the extent of the destruction: out of 1227 nationally protected residential buildings, 38 were destroyed, 222 were severely damaged, and 460 were rendered unusable.

Over the years, these destructions, combined with other factors such as the lack of guidance or regulations for interventions on ancient houses and unclear communication regarding protective restrictions when purchasing property, have led to the current situation. Despite the efforts of many international organizations, residents, and Armenian institutions, the historic part of the city remains underutilized, with extensive areas experiencing severe decay, affecting not only the buildings but also the open spaces and public services.

Some of the buildings damaged by the earthquake have been abandoned, while others have been demolished over the years, leaving behind large urban voids. The remaining historic fabric, when still inhabited, undergoes daily, albeit small, but significant transformations by the residents. New volumes are often added using temporary materials to expand living space, disregarding both the architectural character of the historic building and its capacity to bear additional structural loads.

Original windows, with their unique closing systems and double-leaf design to protect against the harsh winter climate, are being replaced with new plastic or aluminum fixtures. Additional elements, such as canopies and awnings, have been installed to highlight entrances and provide shelter from rain or snow.

Although there is no comprehensive inventory of the current state of Gyumri's architectural heritage, it can be estimated that over 50% of its historic buildings suffer from one or more of these issues. Demolition and abandonment are concentrated in certain areas of the historic center, while transformation projects and new additions are widespread throughout the ancient core.

These processes risk eroding Kumayri's identity and fundamentally altering a city that has preserved its architectural heritage and urban landscape for centuries.

In addition to threats to its cultural heritage, Gyumri faces several social problems, including a lack of job opportunities and low pension incomes for the elderly population, inadequate healthcare and social services, and the poor living conditions of historic center houses occupied by this vulnerable demographic.

Even today, the "*domik*" (temporary containers erected after the earthquake to house those who lost their homes) remain, surrounded by newly built structures often made of temporary materials and suffering from inadequate sanitary conditions, leading to overcrowding. Many streets, particularly those running through residential blocks, lack proper paving and lighting.

In some areas of the historic city, the ruins of destroyed buildings are still visible, as well as large, overgrown urban voids. Only the main square and the immediately adjacent streets appear to have received some intervention, with improvements including well-maintained open spaces, pedestrianization of certain sections, carefully done pavements, and an effective lighting.

These areas serve as a "showcase" for the city, where people stroll or linger in the newly opened establishments. Even in these sections, which have attracted some international investors, the historic buildings are at risk—not due to abandonment or deterioration, but because of ongoing alterations. Similar to the capital, Yerevan, contemporary buildings of mediocre architectural quality and completely out of context are beginning to appear. Even worse, there are instances where only the historic facades are preserved, "attached" to new concrete structures.

Further south, in the area of the old covered market, the situation is different. Only the ground floors of historic buildings are in use, with continuous rows of shops selling inexpensive goods and artisanal products. In the streets, a spontaneous market of fruits and vegetables highlights the scarcity of structured employment opportunities for residents. To supplement their income or pension, many people sell produce from their gardens and small plots of land.

There is no doubt that Gyumri's revival must involve a strategic national planning effort capable of restoring the city's role within the country. This would attract new investments to develop a productive fabric that generates stable employment for residents, as well as secure housing and essential services such as hospitals and schools, to prevent further emigration to the capital or abroad.

In this process, preserving the extensive historic building fabric that still remains is crucial. It makes Gyumri a city of significant cultural interest and importance on an international scale. The Armenian government has historically aimed to make Gyumri the cultural capital of the country and an international tourist center, potentially benefiting the surrounding region as well. A new road under construction is expected to connect the city more efficiently to Yerevan. Although it may take time to complete, this infrastructure could stimulate the development of new businesses and enhance tourism.

Today, however, the situation remains quite different. The much-desired development of tourism in the city has not yet taken off, particularly medium-stay tourism that could bring substantial economic benefits. While there is a noticeable influx of tourists visiting the city's central areas, especially during the summer months, most of these visitors only stay for a single day. Consequently, the potential economic benefits that tourism could provide to the city and its residents are minimal.

Improving the population's living conditions and the city's overall livability can foster lasting and sustainable tourism, complemented by new and engaging cultural activities and itineraries. Kumayri possesses undeniable potential, drawing from its extensive and rich heritage, which could help make the city appealing once again to both returning and new residents, as well as tourists. New job opportunities could emerge from leveraging the remaining artisanal skills, such as the production of ceramics, textiles, and locally sourced food products.

However, this potential can only be realized if the destruction and transformation of the historic heritage are halted. Meetings with local authorities and stakeholders have proven very useful in identifying their needs. Besides highlighting the aforementioned issues, these meetings revealed the need for precise and specific regulations for the historic area of the city to balance the preservation of its historic fabric with the improvement of residents' living conditions. In particular:

- The need for legislation regulating planning (planning instruments).
- The need for building regulations for different zones: rules for buildings, streets, and public spaces (gardens, squares).
- The need for detailed and specific zoning for other parts of the city.
- The need for clear definitions of constraints on buildings and communication with their owners.
- The need for guidelines for maintenance interventions aimed at both professionals and residents.

Based on the identified potential and challenges, as well as the needs of local institutions and residents, proposals and tools for the protection of Gyumri's historic city

Fig. 4.5 Additional storey onto a historic building. The increase in the need for housing space due to family growth or the cohabitation of multiple households for economic reasons results in the addition of new volumes constructed informally with temporary materials and without adhering to the architectural principles of the building (*Photo* F. Vigotti, 2022)

center have been developed, in coordination with the Urban Development Committee of the Republic of Armenia and its policies for revitalizing this extensive planned historic center [4].

4.3 Illustrated Directory of Primary Problems of Decay of Kumayri and Its Historic Buildings

The city of Gyumri, and the oldest part in particular, shows a varied series of transformations of a utilitarian character carried out spontaneously by the inhabitants, as well as restoration interventions that do not respect the historical characteristics of the building. Although with different characters and shapes, linked to the specificity of the buildings, their materials, and local construction techniques, these problems share Gyumri to many cities of emerging countries, where the so-called historic center has not yet undergone gentrification and is, instead, inhabited by the economically most fragile population. This photographic collection aims to illustrate the main problems detected in the Armenian city (Figs. 4.5, 4.6, 4.7, 4.8, 4.9, 4.10, 4.11 and 4.12) but also a more general identification of those spontaneous and nonspontaneous transformations that must be considered when addressing the issue of urban conservation in emerging countries.

The systematic study of these changes will help to understand the new needs of the inhabitants and suggest interesting ideas for possible strategies for their regulation.

4.4 Towards a New Identity

The revitalization of Kumayri requires more coordinated actions, taking a global perspective on the historic center. As the era of heavy industry concludes, Gyumri should position itself as a hub for emerging economies and align with global trends in

Fig. 4.6 Temporary elements and superstructures constructed with provisional materials and without a design that considers the architectural features and structural resistance of the building (*Photo* F. Vigotti, 2022)

Fig. 4.7 New elements projecting on a road section to emphasize the entrance or protect it from weather. The absence of regulations and defined rules results in the appropriation of public land for private use (*Photos* F. Vigotti, 2022)

ecological and digital transitions. In this context, three key areas could be particularly promising for Gyumri:

- Cultural City
- Smart and Digital City
- Organic Agricultural City.

Implementing these three functions concurrently could spark a revival for Gyumri, establishing it as a pivotal city both within the country and in the broader geographical

Fig. 4.8 The gas distribution network is still above ground. Often the utility network runs visibly in the historic centers of emerging countries (*Photo* F. Vigotti, 2022)

Fig. 4.9 Historic buildings ruined by the earthquake. In the historic center, in the absence of public funding or incentives to support residents often in difficult economic conditions, buildings remain in a state of ruin, abandoned because their restoration is deemed too costly (*Photo* F. Vigotti, 2022)

context. This approach would also breathe new life into its extensive and unique historic center by introducing new functions.

Given the exceptional quality of its historic center, positioning Gyumri as a cultural city is the most natural and fitting direction.

A city of culture encompasses more than just tourism; it also involves developing a system of functions tied to cultural activities. To achieve this, it would be necessary to:

- **Enhance the Museum System**: Improve existing museums, establish new ones, and organize the entire historic center as an open-air museum. Digital itineraries could be created to highlight and explain the main monuments and the unique historic buildings that define Kumayri. Using QR codes, these digital routes

Fig. 4.10 Unpaved pedestrian connections and buildings of poor quality built after the earthquake. The lack of funds from the public administration leads to prioritizing, in the creation of services and spaces for the resident population, certain parts of the city considered more "representative", at the expense of the inner parts of the old town (*Photo* F. Vigotti, 2022)

Fig. 4.11 Renovation of a historic building: total demolition of the internal structures and replacement by reinforced concrete pillars. These interventions, quite common even in the capital Yerevan, use what remains of the historical facades to "embellish" contemporary buildings, considering such works equally effective for public image and less costly than a careful and respectful conservation project (*Photo* S. Pistidda, 2022)

could be augmented with small installations at strategic points throughout the historic center. Virtual reality could also be utilized to reconstruct and illustrate the historical phases of Gyumri, offering immersive experiences of the city's past. These itineraries will attract tourists and raise the local population's awareness of heritage preservation.

- **Implement Cultural Activities**: Host exhibitions, theatre performances, and concerts. These events could be organized in existing theatres, public squares, and

Fig. 4.12 *Domik* was built as an emergency house after the earthquake and is still present in some areas, including the historic center core areas (*Photo* S. Pistidda, 2022)

streets during the spring and summer. Additionally, some historic buildings could be used for chamber music concerts, experimental theatre, and small exhibitions.

- **Encourage Artists' Return**: Allocate historic residential buildings for artists to live, work, and exhibit their creations. After many years, some artists who had left Gyumri due to the 1980s earthquake have returned, and providing dedicated spaces for them could foster a vibrant cultural scene.
- **Promote Artistic Handicrafts**: Support the development of artisanal crafts such as jewelry, ceramics, and carpets. This initiative could create job opportunities for young people and women in skilled trades and attract tourism. A "home-workshop" project, where artisans can live, work, and sell their products, could be particularly effective. Many historic houses are suitable for this purpose; the ground floors, without requiring new openings or storefronts, would be ideal for retail spaces.

The idea of promoting Gyumri as a smart and digital city stems from several existing or upcoming initiatives.

Foremost among these is the TUMO Center for Creative Technologies, which offers experimental, advanced, and self-directed learning in fields such as Animation, Game Development, Filmmaking, Web Development, Graphic Design, 3D Modeling, and more. This cutting-edge, entirely free school, made possible by a group of donors, has the potential to attract young people from the region and across the country, fostering the development of new professions.

The second initiative is the planned construction of a futuristic Technopark in the semi-central area of Gyumri. While the actual visual and prospective impact on the historic center needs to be evaluated, the Technopark could create new job opportunities in the city and the region. Its presence could be particularly beneficial if it manages to establish synergies with the historic center by decentralizing certain functions. For example, integrating hospitality areas, meeting rooms, and smart-working spaces within historic buildings could offer dual benefits: limiting the expansion of new construction in the Technopark, thus preserving land, and repurposing currently abandoned historic buildings. Connected to these two hubs could be the development of sustainable mobility, which could position Gyumri as a pilot city in this area in Armenia.

Lastly, there is the potential to establish Gyumri as a center for organic agriculture. Many small farmers in Gyumri grow vegetables and fruit in their gardens and sell them in the streets adjacent to the covered market, which is partially unused due to high space rental costs. Moreover, the region retains an agricultural vocation, currently characterized by local, organic produce, more out of necessity than choice.

Preparing the population for organic farming and implementing spaces dedicated to it would present an opportunity to create new jobs. An agricultural school focused on sustainable agriculture could further stimulate this activity, especially given the large open areas on the outskirts of the city. The school could become a national reference point and attract young students to Gyumri. One of the many unused buildings in the city center could serve as an ideal location.

A future master plan for Gyumri, which is expected to be developed in the near term, could incorporate the strategic functions described above. These functions emerged from a study of the city's challenges and opportunities, particularly concerning the potential uses of many currently abandoned historic buildings, with the aim of initiating the much-needed revitalization of the historic center.

4.5 Planning and Managing Kumayri. A Strategic Framework for Heritage Preservation

The preservation of the historic part of Gyumri and its buildings should be entrusted not only to the numerous existing national protection regulations and the resumption of activities by the group of experts active until the 1980s, but also, and most importantly, to the drafting of a Strategic Plan. This plan should promote improvements in housing conditions, adapt services for residents, foster new economic activities, and ensure the preservation of the widespread architectural heritage.

Gyumri, a rare example of integral preservation of its historic core, has so far been preserved precisely due to its lack of "development". Paradoxically, the devastating earthquake that struck and severely damaged the city, leaving it partially depopulated and deprived of economic activities, has ensured its preservation, as if frozen in time for half a century. Tourism, currently more aspirational than actual, could rapidly alter the state of the city, potentially transforming it into an idealized and unrealistic image tailored to visitors. Signs of this trend are already appearing on the main street of the historic city. Therefore, balancing tourism development with residential needs must be a priority.

A crucial aspect is improving the livability of older residents and attracting new ones to prevent the abandonment of Kumayri's historic center.

The General Plan should include initiatives to promote social housing, attract young people from other areas, and encourage the younger generation to stay in Kumayri, thereby contributing to its preservation and heritage transmission. The concurrent development of artisan activities, with a focus on quality products and organic agricultural improvements, can further support the repopulation process.

For Gyumri's revitalization to succeed and be grounded in the preservation of its architectural heritage, it is essential to engage the local population and help them see living in a historic center as a privilege rather than a source of inconvenience and discomfort.

However, necessary interventions could profoundly alter, if not destroy, the architectural heritage. To mitigate this risk, it may be useful to develop a set of material and moral incentives to engage residents in preserving the historic centers. For instance, tax reductions under specific conditions—such as using particular materials or adhering to preservation guidelines—could be implemented. Reducing or eliminating property taxes, land taxes, or other related taxes could encourage property owners to support preservation efforts.

Interventions that comply with the guidelines established by the municipal administration for historic buildings could be eligible for an annual special award, accompanied by a public event for discussion and sharing. A commemorative plaque on the building's façade could recognize the award received, potentially increasing the value of the involved properties.

Yet, the involvement of the local population and their engagement in preserving the historic center will not be sufficient without an appropriate planning tool to regulate interventions in the historic city. To support the creation of a plan that aligns with Gyumri's character and moves beyond the zoning-based approach currently used in Armenian planning tools, even for historic centers, several support mechanisms have been proposed. These mechanisms, varying in nature, role, and purpose, aim collectively to contribute to the development, approval, and management of the plan.

4.5.1 The Creation of a Planning Review Board

Establishing a Planning Review Board is a crucial aspect of plan management, ensuring both the quality of the process and the results. The Board could be established by municipal resolution and granted powers similar to those of the office responsible for preserving the historic center until 1988.

The economic resources required for the Board's operations could be minimal, with members potentially working on a voluntary basis initially, akin to the Boards in English garden cities or the "landscape commissions" in Italy, which offer non-binding opinions on specific interventions.

The Board will specifically oversee interventions to ensure urban decorum and compliance with the general objectives of the Plan. Comprising members with diverse expertise, skills, and backgrounds and incorporating a periodic rotation mechanism, the Board can foster productive and transparent discussions about the impacts of the Plan's implementation.

The Planning Review Board will focus on preserving areas of cultural significance and ensuring that buildings, structures, signs, and other developments adhere to high standards of taste and design. It will ensure that these elements are harmonious with surrounding developments and contribute to maintaining Kumayri's vision and reputation as a place of historic importance, spaciousness, and quality.

The Board will aim to enhance public awareness of historic, cultural, architectural, and archaeological preservation by:

- Protecting buildings identified and listed as historic, cultural, architectural, or archaeological resources.
- Holding hearings and recommending to the Municipality the designation of specific structures and properties as Historic Structures.
- Enforcing design guidelines for the Historic Overlay Districts.
- Reviewing public works and utility projects proposed within or adjacent to Historic Overlay Districts.
- Reviewing and approving projects to ensure compliance with design guidelines.

To establish the Planning Review Board, it is essential to define its composition, terms of office, recruitment methods, and operational model. The Board should meet regularly, at least once a month, to review ongoing projects.

The Board is suggested to have a diverse composition, including representatives from institutions as well as individuals with specific skills. An interdisciplinary approach is recommended to cover various aspects, including architects, engineers, sociologists, restorers, urban planners, economists, and others. Each member should be assigned a specific role within the decision-making process and according to the steps to be followed.

Agreeing with UDC on the urgency of the issue, the following composition for the Board is proposed:

- One representative from the Municipality (either the Mayor or another member of the Council).
- One representative from the Ministry of Culture.
- One representative from UDC.
- Three experts selected by UDC and the Ministry of Culture from among relevant academics or professionals.
- Two or three community members elected by the inhabitants, preferably from associations already active in Kumayri's social and cultural spheres.
- One or two external consultants with recognized expertise in the field to provide additional balance.

Diverse recruitment methods are important to ensure transparency. Some members may be selected based on their qualifications by a jury of local and national experts. An open call for applications could be issued to gather candidates, while other members could be proposed and elected by the local community through an open consultation process.

The Board should serve a term of two to three years to ensure member turnover, with clear term limits established for all decision-making members to maintain effective governance and fresh perspectives.

Regarding the operating model, an "open society" based on a cooperative model could attract new investors and financing. While this proposal is under discussion, it is crucial to define the society's terms, membership requirements, and all legal aspects. A model with mixed participation from residents and private investors—where residents contribute through taxes and investors provide company shares—can offer economic sustainability to the initiative.

A dedicated website for the "Kumayri Reserve", linked to major social media platforms for rapid communication, is recommended to manage the entire process. The website should offer information on the site's history, ongoing projects, details about the Reference Board, guidelines for interventions on historic buildings, and legal documentation. This platform will enhance public engagement and transparency throughout the process.

4.5.2 The Definition of Priorities in Interventions

To be effective in its implementation, the Plan should incorporate certain priority interventions within its strategy to restore the city's attractiveness, attract investors (including foreign investors, tourists, and new residents), and thereby initiate a virtuous cycle of revitalization.

In particular, the following priorities for the historic center should be considered:

- Providing guidelines for the urgent removal of asbestos roofs, which are widespread in many buildings in Kumayri and pose significant health risks.
- Conducting a census of the remaining *domik* structures and assessing their current use.
- Conducting a census of building ownership to determine which buildings are privately owned, publicly owned, or unclaimed. This information would be instrumental in planning strategies for the historic center and selecting appropriate public uses.
- Conducting a census of the population's living conditions in the historic buildings.
- Verifying the conditions and legality of new buildings that have been constructed within the blocks.
- Regulating and, if possible, halting new constructions and the demolition of old buildings, even if they are in a ruined state, until the new master plan is approved.
- Providing social housing in selected parts of the historic center. Organizing a social housing project in unused historic buildings will be essential for improving the living conditions of the population in the historic center. Many abandoned and degraded blocks are located just beyond the main square and the two pedestrian streets. These sites, as they stand, could struggle to attract tourists, new inhabitants, or investors.

To revitalize a historic center like Gyumri, where building conditions are often problematic and the historic area is quite extensive, it may be necessary to carry out interventions simultaneously in different parts of the city.

It would be advisable to select two or more pilot zones of limited size and fully rehabilitate them, including both buildings and outdoor spaces, restoring residential use and introducing new attractive functions and public services. These "fires" could serve as pilot projects, demonstrating proper rehabilitation techniques for historic buildings and acting as catalysts for the revitalization of other areas, driven by residents organized in cooperatives or by private investors.

For the identification of these pilot areas, studies on Kumayri and meetings with national and local stakeholders have highlighted some priority issues. These could serve as the focus of pilot projects for the regeneration of the historic center.

Specifically:

1. Social housing projects (Fig. 4.13).
2. Creating green areas and footpaths in crowded courtyards (Fig. 4.14).
3. Designing a greenway above the buried river between Abovyan and Shaumyan streets (Fig. 4.15).
4. Redesigning the main square (Vartananz Square), the entrance to the historic city (Teriyan—Tigranyan streets), and Victory Park (Fig. 4.16).

Fig. 4.13 Social housing projects. In addition to the necessary housing interventions described above, such as identifying two or more significant areas, certain parts of the city could serve as focal points for pilot redevelopment projects. These zones are crucial for the city's overall revitalization, impacting both its public image and the livability of residents and visitors. Specifically, areas east of Liberty Park Boulevard and near the main square could be ideal for launching these pilot projects focused on housing interventions (*Photos* By authors, 2022)

Fig. 4.14 Creating green areas and footpaths in densely built courtyards. Many of the courtyards behind the buildings and areas once built up have become vacant due to the earthquake's destruction. In recent years, these spaces have been filled with temporary structures that were built without consideration for the historical grid of the area. Once the issue of relocating the inhabitants of these temporary houses is addressed, these spaces could be partially cleared of the more dilapidated new buildings to create green areas for the use of residents from various blocks. The development of pedestrian pathways could further enhance these areas, making the interior of the blocks more vibrant and connecting different parts of the city through car-free routes. Like housing interventions, one or more pilot areas could be selected for the medium-term implementation of these green spaces and pedestrian pathways to test and refine the effectiveness of the intervention (*Photo* By authors, 2022)

Fig. 4.15 Designing a greenway above the buried river (between Abovyan and Shaumyan streets) presents an opportunity to enhance the urban landscape while preserving historic elements. This area is ideal for creating a tree-lined pedestrian path, offering a serene and scenic route that also opens up views of the Cathedral, a key landmark in Gyumri. In addition to developing the greenway, it would be necessary to redesign the façades of the buildings facing this path on both sides (*Photo* By authors, 2022)

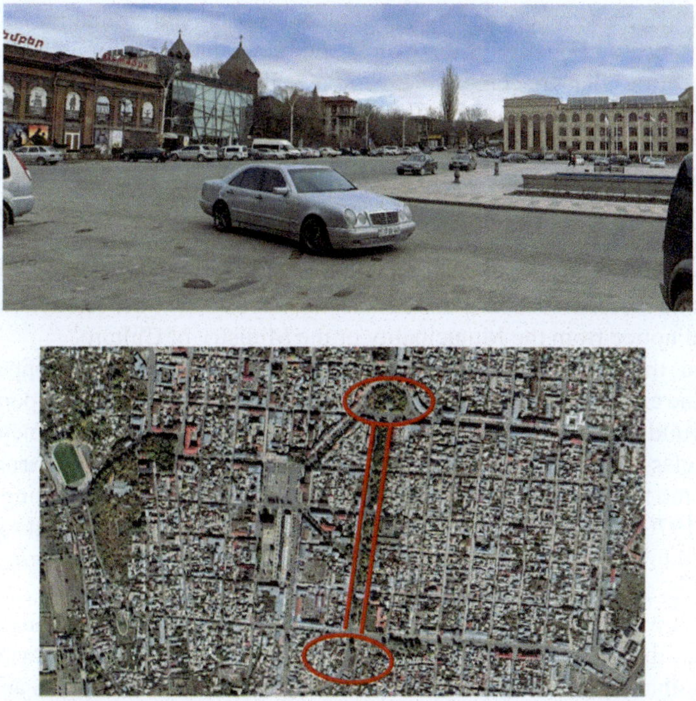

Fig. 4.16 Redesigning Vartananz Square, the entrance to the historic city (Teriyan—Tigranyan streets), and Victory Park is essential for enhancing the city's core and making it more attractive and functional. Vartananz Square, the heart of the city, requires a thoughtful redesign to increase its appeal. Without altering the design that reflects an important historic phase of the city, improvements could include the introduction of trees and seating areas, the reorganization and partial elimination of parking spaces, and the redesign of traffic flow to expand pedestrian areas. Similarly, the entrance to the historic city and Victory Park, which features a tree-lined avenue of some interest, needs to be reimagined to play a more prominent role in the urban landscape. This could involve not only improving the avenue itself but also maintaining and, in some cases, redesigning the building facades along both sides. Additionally, the monuments within the central part of the avenue should be carefully restored or redesigned to enhance the avenue's significance and aesthetic coherence

4.5.3 Special Tools and Regulations for the Preservation of the Built Heritage and the Historic City

The interviews with stakeholders conducted in Gyumri identified a significant challenge in preserving historic buildings: the lack of awareness among property owners about the preservation restrictions on their buildings. This lack of knowledge often leads to indifferent or inappropriate interventions that do not respect the historical value of the structures. Once such interventions have been made, it becomes very difficult for the authorities to reverse them, nor would it be practical to do so. Therefore, it is necessary to implement a series of actions aimed at preventing such behavior.

First, raising awareness among the population is crucial so that they understand the importance and quality of the historic buildings they inhabit and the necessity for interventions that are compatible with the building's historical significance. This can be achieved through conferences, workshops, public discussions, incentives for exemplary restoration work, and providing guidelines to residents for carrying out compatible and high-quality renovations. However, these awareness-raising initiatives should be supported by concrete procedures. For instance, when a property is transferred to a new owner, the details of the existing restrictions on the building could be included in the purchase deeds. If the ownership remains unchanged, the owners (and residents) should be formally informed about the restrictions through an official notice from the Municipality or the Ministry of Culture.

Second, the development of a new Master Plan could serve as an opportunity to reassess the existing restrictions on historic buildings. Currently, only a portion of the historic buildings in Kumayri are designated as "Architectural Monuments," while most are classified according to the number of floors. It may be more appropriate to extend some form of protection to all buildings in the historic center dating back to at least the 1970s, even if this protection is less restrictive than the current regulations. These buildings could be classified as *contributing to the urban identity*, ensuring their preservation and integration within the city's historical context.

Third, it would be advisable to include a series of mechanisms within the Plan to ensure that the projects submitted genuinely preserve the historical integrity of the buildings, thereby avoiding heavy restructuring of both internal spaces and external facades. These mechanisms could provide guidelines for both large-scale interventions, which, given the conditions of certain areas of the city, might attract significant investors, and for smaller property owners. In the case of large-scale interventions, the Plan could outline the boundaries of specific areas identifiable as minimal units for a cohesive intervention.

These large-scale interventions should be subject to specific detailed plans proposed by private investors. This approach allows for the advance verification of the architectural quality of the interventions, ensuring they align with the broader urban planning objectives and comply with the strategies established by public authorities.

Defining a minimum quality standard for the documents required for the submission of projects involving buildings designated as "Architectural Monuments" or as "Buildings Characterizing Urban Identity" could provide an additional criterion for assessing the suitability of the projects. This would apply both to large-scale interventions and to individual buildings.

The proposed list of documents could be based on two principles. The first principle pertains to the type of protection constraint on the buildings, which may vary from more stringent to less stringent. The second principle relates to the conservation status of the building, which determines the required level of intervention in terms of the extent of work needed.

As a general rule for Gyumri, three different levels of document lists could be considered for submission at the time of project presentation:

- **List 1**: For buildings not classified as "architectural monuments", covering minor interventions.
- **List 2**: For buildings listed as "architectural monuments".
- **List 3**: For specific interventions affecting both architectural and non-architectural monuments.

In all cases, these lists, which can also serve as checklists to assist professionals and residents in project submission, will direct them to conduct studies on the conservation status, document the current state of the building with photographs, and perform surveys and drawings at a scale appropriate to the planned intervention. Proper submission of study and project documents will facilitate the reviewing authority's task and, consequently, streamline the approval process.

Another issue that the Plan should address concerns the insertion of new buildings into the urban voids left by the earthquake within the historic fabric.

First, in Kumayri, the inclusion of contemporary architecture should be preceded by a careful assessment and without uncontrolled demolition of historic buildings. These buildings could be preserved through appropriate conservation projects.

In exceptional cases where the destruction from the earthquake and years of neglect have led to the disappearance of historic buildings, the Plan may suggest that new buildings should respect certain minimal characteristics. These characteristics should include being recognizable as contemporary architecture (i.e., avoiding reconstruction in a historic style or imitation of past buildings) while relating to the historic context.

While it is challenging to prescribe the architectural language of new buildings (and it wouldn't be fair to do so), the Plan can include specific guidelines regarding their relationship with the historic context. These guidelines should address aspects such as:

- Respecting the heights of surrounding buildings
- Aligning with the street alignment of surrounding buildings
- Considering the frequency of openings in neighboring buildings

The building materials must also be considered very carefully to avoid creating shadows or reflections that could distort the historic context.

The design of new buildings, along with all necessary documentation to understand their architecture and construction details, must be subject to analysis and approval by a committee of experts, potentially including international professionals.

Lastly, the Plan should address two other fundamental issues for the preservation of Gyumri's historic fabric. The first concerns the city's open spaces.

The outdoor spaces are an essential part of the historic city, as are the buildings that compose it. They characterize the urban shape and make the city unique.

In Kumayri, a planned historic city of founding, their preservation, both in the ratio of full voids and the materials used to make them, takes on particular importance.

The Urban Plan should carefully address this and establish specific regulations for the preservation and maintenance of the characteristic "chequerboard" layout. Special attention must be given to preserving the existing paving, as it is a defining feature of the historic center. Even with similar materials and designs, replacement of the original paving is strongly discouraged, as contemporary stone processing techniques produce results that differ significantly from the traditional methods used in the past.

The second issue concerns the so-called "urban decorum". The recent opening of new businesses near the main square and the variety of signs and other elements introduced by different owners highlights the need for the new Plan to include an annex with specific guidelines for urban furniture.

The decorum plan can be organized by detailing all urban elements through a series of selected examples, providing guidelines for their appropriate use and placement.

The Plan for a historic city like Gyumri, which has been preserved but may face significant transformations in the coming decades due to modernization pressures, will need to go beyond the scale of the master plan. It should delve into the specifics of detailed plans and develop tools for the conservation and management of historic built heritage. This detailed planning must be tailored to the unique characteristics and conservation status of the historic city.

This issue is relevant to many historic cities in emerging countries, which will similarly face challenges related to modernization and preservation.

References

1. Vv AA (2013) Gyumri. Eldorado Publishing House, Gyumri
2. Arakelian S, Martirosyan L, Mkrtchyan R (2014) Conservation and reuse project of buildings and open spaces in Gyumri. Specializing Master thesis, Master in "Conservation design. Training in Armenia in the restoration. A project for development cooperation", YSUAC-State University of Architecture and Construction of Yerevan, 2011–2014
3. Barsegyan L (s.d) Report on residential buildings of Kumayri Museum Reserve, National Archive of Armenia
4. Bold J, Pickard R (2018) An integrated approach to cultural heritage: The Council of Europe's Technical Cooperation and Consultancy program, Council of Europe.
5. Dolukhanian LK (1980) On the soviet Armenian architecture of the 20th century. S.e., Yerevan (in Russian)
6. Griffin MJ, Bragagnoio LJ, Yanev PI (1991) The December 7, 1988, Armenia earthquake effects on selected power, industrial, and commercial facilities, prepared for Electric Power Research Institute, Palo Alto, California
7. Harutyunyan M (2020) Gyumri: architectural and design observations. Pol J Sci 24:5–12
8. Israyelian R (1982) About the architecture of Leninakan. Articles, research, reviews. Institute of Art by the Academy of Science of Armenian SSR, Yerevan
9. Ivanov A (2020) Armenian city of Gyumri as a phenomenon of living vernacular urban environment. In: The international archives of the photogrammetry, remote sensing and spatial information sciences, international conference on vernacular architecture in world heritage sites, risks and new technologies, vol XLIV-M-1-2020
10. Lachikyan L (2012) Gyumri—the salt of Armenia. In: Civilnet, 9 Dec 2012
11. Lynch HFB (1901) Armenia, travels and studies. Longmans, Green and Company, Londan
12. Matevosyan S (1985) Gyumri's national architecture. S.e., Yerevan (in Armenian and Russian)
13. Qalashyan G (1982) Kumayri: state historical and architectural reserve museum. S.e., Yerevan
14. Qalashyan SH et al (2021) Microregional complex urban planning development plan and zoning of Shirak region settlements (Gyumri and rural settlements). Armproject OJSC, Yerevan

15. Qalashyan SH, Aloyan AA, Petrosyan NG et al (2001) The zoning plan of Gyumri historic center. Armproject OJSC, Yerevan
16. Saxutyan ST (1989) The Leninakan Monuments, report. National Archive of Armenia, 6 January (in Armenian)
17. Villari L (1906) Fire and sword in the Caucasus. The Fisher Unwin, Londra—James Pott and Company, New York
18. Von Thielmann MFGF (1875) Journey in the Caucasus, Persia, and Turkey in Asia. John Murray, Londra

Chapter 5
Building a "Heritage Community" in Gyumri Through a People-Based Maintenance

Abstract The document presented below derives, with a few revisions to clarify the meaning in the context of the book, from the research report delivered to the clients of the Asian Development Bank and the Urban Development Committee of the Armenian Government. It consists of guidelines for the maintenance of historic buildings intended mainly for inhabitants and unskilled workers. The choice to propose them maintaining both the simple language and the graphic layout that distinguished the document delivered, is motivated by the intention to propose them as possible methodological tools addressing the theme of the existing city in emerging contexts. Although specifically designed for the city of Gyumri and not exportable uncritically to other realities, they may suggest methods to sensitize the inhabitants to the fact that they live in a cultural heritage intended as a complex formed by the city and its individual buildings. The awareness of the uniqueness and irreproducibility of this Heritage implies, consequently, a more respectful intervention in existing buildings. In contexts where self-construction also concerns the architectural heritage, easy indications are necessary to carry out the necessary interventions to provide the buildings with greater living comfort.

Keywords Best practices · Restoration criteria · Guidelines for historic buildings · Heritage-based approach

5.1 Introduction

This chapter addresses "practical" indications for the conservation of historic buildings [1].

These practical indications should not be considered as manual with standardized intervention techniques [14]. Site-specific studies and knowledge are fundamental to orient the choices. On the contrary, they are a sort of guideline helpful to residents and workers to understand the direct link between the problems and solutions. A similar proposal of guidelines was first developed by the authors for another city in an emerging country: Multan in Pakistan [4].

M. Giambruno et al., *Heritage-Based Tools for the Sustainable Urban Development of Historic Centers*, PoliMI SpringerBriefs, https://doi.org/10.1007/978-3-031-75259-9_5

The initial reflections were entirely analogous to the case of Gyumri:

- The historic center was vast and complex.
- The buildings highlighted numerous decay phenomena, instability, and spontaneous additions incoherent with the historic architecture.
- The inhabitants had very modest economic resources and they carried out most interventions in self-construction without awareness of inhabiting a historic building.
- The labor force had no specialization in restoration works.

The possible choices were two. Either reaffirm that the conservation of existing buildings requires extensive research studies, a thorough and detailed project, and qualified specialized companies or attempt to provide the minimum necessary knowledge to residents and non-specialized artisans. The second path was chosen, not without deep theoretical reflection and some doubts about the effectiveness of the tool. The issue is not so different from what is daily faced in Countries where the culture of Built Heritage Conservation has been debated and addressed for years.

If every historic building can be considered a "monument", identical study and intervention methods should be applied to both exceptional objects and common houses. However, for the latter, the issue becomes complex because they are more numerous, inhabited, and often not considered heritage. It is not uncommon to witness their demolition or interventions that, without the slightest preliminary study, replace parts because they are considered a priori not resistant or too degraded. Similarly, there is no obligation to employ specialized labor for interventions on historic buildings. If this happens in the so-called "developed" countries, the phenomenon is even more pronounced in "emerging" ones where the issues above are compounded by an even more significant lack of preparation among designers, a labor force accustomed to working only on new constructions, and the economic fragility of the population inhabiting the historic areas of the city, often the economically poorest [3].

Entrusting the conservation of the historic city of Multan, as well as Gyumri, solely to the preparation of designers and labor would mean losing most of the historical evidence currently preserved. Hence, the choice to develop user-friendly guidelines illustrated with numerous images and graphics that are easy to understand.

The guidelines for Gyumri are composed of three sections.

The first deals with general themes related to the problems and conservation of historic buildings, suggesting general criteria for their restoration and reuse. Moreover, it illustrates good and bad practices to provide residents with a general framework of the main mistakes that can be made during the intervention.

The second section collects some practical factsheets suggesting intervention methods to solve the main degradation issues identified during the site surveys. Residents and non-specialized laborers can effectively draw inspiration from these in carrying out the work.

The third section offers a checklist to guide residents in recognizing the main phenomena of decay and instability affecting the building they live in and to assist them in choosing the most appropriate intervention.

5.2 Recommended Practices. General Criteria for Conservation and New Uses Inclusion in Historic Buildings

Preserving the characters and materials of the historic buildings is the prerequisite that orients the guidelines. REPAIRING must always be the first action, avoiding replacements and alterations of existing buildings as much as possible. Replacements (limited and carefully evaluated) are allowed only in case of irreversible decay. The preservation approach orients the attention to the SUBSTANCE of the buildings (and not to their image!): materials, finishes, decorations, and construction techniques. The uniqueness of a building is also given by its stratification process: additions and transformations must be carefully investigated to understand their role in the changing identity of the building [9].

5.2.1 Basic Requirements of the Conservation Interventions

The preservation works on the existing buildings, as well as new insertions, have to be carefully evaluated with specific investigations, and, in general, they have to satisfy requirements of COMPATIBILITY (of materials used but also new uses), DURABILITY (the intervention must guarantee a good duration over time), NON-INVASIVENESS and RECOGNIZABILITY (the new materials and structures have to be clearly visible and not imitate previous styles, according to the international standards). To identify new uses, the character of the buildings, their spatial relations, and their current distribution must be taken into consideration to satisfy the "vocation of use" of the existing built environment as much as possible (Figs. 5.1, 5.2, 5.3, 5.4, and 5.5). This approach allows us to limit demolitions and irreversible transformations as much as possible [2, 8].

5.2.2 General Suggestions

Yes! Conversion of temporary structures to new uses

No! Alteration of features

Yes! Check the drainage system

Fig. 5.1 In Gyumri there still are several unused free areas or resulting spaces, abandoned and degraded. These parts of the city can be the opportunity to build new connections, open perspectives, and give space to the built environment and new green and pedestrian areas to the inhabitants (*Photos* S. Pistidda, 2022)

Fig. 5.2 The electrical and gas systems must be subjected to a systematic rethink, avoiding random interventions as much as possible but inserting them into an overall project. The identification of suitable external conduits must be made by considering the design of the façade and its decorative elements, exploiting already existing cavities (moldings, eaves, frames) as much as possible, and paying attention to the colors (*Photos* S. Pistidda, 2022)

5.2.3 Guidelines for Historic Buildings Seismic Improvement

Several buildings damaged by the earthquake of 1988 are still in awful condition today, highlighting severe structural problems (Figs. 5.6 and 5.7). As a preliminary measure to avoid collapsing, the damaged masonry must be stabilized with consolidation works before undertaking appropriate preservation works. As general indications, especially in the restoration works, it is suggested to:

- Make an initial assessment of the structural decay and identify causes of instability to define the level of risk of the building: connections between the elements that

Fig. 5.3 The "*domik*", the temporary structures designed to respond to the emergency immediately after the earthquake of 1988, continue to be used today. An organic map of all the existing systems and their actual uses is necessary to abandon the current residential functions (not suitable for the current needs) and to plan the removal of the most degraded ones or, where possible, their conversion into other functions for the entire community (spaces for artists, social centers, spaces for reading) (*Photos* S. Pistidda, 2022)

Fig. 5.4 Several existing roofs are made of harmful materials such as asbestos. These materials must be removed and replaced with new ones. Moreover, new roofing structures are frequently made in different colors and types of metal. This discontinuous treatment makes perceiving the urban fronts as unitary organisms difficult. Detailed suggestions on roofing works (materials, colors, construction techniques to be used) would be necessary to make them more homogeneous (*Photos* S. Pistidda, 2022)

compose the structure; survey of the crack pattern; survey of displacements and deformations; check of the out of plumb, etc.

- Define the possible causes of instability and their effects: foundation soil, the interaction between soil and structure, and masonry.
- Not enlarge the openings in the façade: the reduction of the wall section may cause the masonry to collapse, besides completely changing the architectural features that define the historical character.

Fig. 5.5 Several historic buildings highlight consistent decay on the fronts due to a lack of protection from environmental agents. One of the first essential actions to prevent it, together with checking the roofing structures, is to protect the masonry from water damage by providing a proper drainage system or implementing the existing ones following the traditional techniques. The new elements should avoid interference with the architecture and decorative features. A periodic check and maintenance of gutters, downspouts, and anchoring systems are also suggested to guarantee efficiency over time (*Photos* S. Pistidda, 2022)

Fig. 5.6 When removing accumulated material, paying close attention to the closer wall is necessary to avoid compromising it. The collapsed parts should be repaired to prevent water infiltration and other structural instability. Grouting the joints between the tuff blocks is fundamental to reinforcing the masonry's mechanical behavior (*Photos* S. Pistidda, 2022)

- Provide adequate roof checking to avoid moisture penetration and other damages. The leaking roofs must not remain unprotected because this condition can accelerate the deterioration of historic materials and structural elements.
- Not increase the height of a historic building.
- When constructing new buildings in the same place as collapsed ones, match the alignment and height of the adjacent buildings.
- Improve the performance of buildings during seismic events by adding structural reinforcements compatible with the existing materials and structures.

Fig. 5.7 The historic buildings that survive only on the exterior façade are strongly vulnerable due to the structural instability generated by the lack of components and the masonries with a consistent loss of elements. These critical situations need urgent provisional interventions to stabilize the masonry and avoid other irreversible collapses (*Photos* F. Vigotti, 2022)

- Check the "box behavior" of the building, avoiding the construction of reinforced concrete curbs [11].

5.2.4 Guidance for the Conservation of Elements Characterizing Historic Buildings (Decorations, Windows, etc.)

Materials, decorative elements, and particular construction techniques represent the uniqueness of Gyumri. Tuff, basalt, sandstone, wood, and metals give rise to unique buildings that define the city's specific character (Figs. 5.8, 5.9, 5.10, 5.11, 5.12, 5.13, and 5.14). The 1988 earthquake left indelible marks, still visible today in many buildings and the emergency "*domik*". While thinking about recovering these buildings from a structural point of view, we must not forget to preserve the elements that compose their uniqueness.

Yes! Materials, decorations, finishes, and construction techniques to be preserved [10]

No! Alteration of features

5.2.5 Guidance for the Conservation of Frames

The traditional material used for the frames of historic buildings is wood, which is now in degraded condition and often replaced with modern frames in PVC or aluminum (Fig. 5.15). The wooden frames should be maintained and repaired

Fig. 5.8 The decorated wooden protruded balconies are typical of some Armenian residential houses. These elements define the outdoor spaces by creating covered passages and rest areas, besides giving three-dimensionality to the urban fronts. For these reasons, they are elements to be preserved (*Photos* S. Pistidda, 2022)

Fig. 5.9 Wooden frames are a characteristic element of the historic buildings in Gyumri. In most cases, due to isolation reasons, a second level of frame is provided in the interiors. For the particularity of the elements, different according to the historic fronts, they are elements that define the uniqueness of the buildings and so to be preserved (*Photo* F. Vigotti, 2022)

or replaced with frames of the same material in case of irreversible degradation. To improve thermal comfort, a second frame can be provided indoors. When not possible, the existing wooden frames can be improved in their characteristics by inserting them between two glass layers. The protection and recovery of the existing wooden frames can be achieved with appropriate surface treatments such as cleaning, paint removal, and re-applicating a protective coating system [13].

Yes! Preservation of the traditional wooden frames

No! Replacement with PVC or aluminum elements

Fig. 5.10 Some of the entrances to the houses are underlined by stairs, decorated shelters, and entrance doors. These elements serve to provide shelter from the rain and frigid temperatures and connect the urban space to the interiors. Hence, they are characteristic elements to be preserved. Sometimes, these elements need a general architectural rethinking to better adapt to the character of the buildings (avoid extraneous materials, colors, and features) (*Photos* S. Pistidda, 2022)

Fig. 5.11 Several facades are enriched with decorative elements, carved in stone, or applied on the surfaces. These elements help create wealth and diversity of architectural elements, so they must be preserved as a unique legacy from the past (*Photos* S. Pistidda, 2022)

Fig. 5.12 Many buildings are altered by incoherent elements, such as added parts to host bathrooms and services, and historic elements are closed by new elements by changing shape, size, and rhythm. These interventions should be avoided to preserve the historic character of the buildings (*Photo* S. Pistidda, 2022)

Fig. 5.13 Many buildings are the result of additions and overlapping that occurred over time. Especially in the finishing of the surfaces, interventions with non-traditional materials such as metal should be avoided. A better result can be achieved by using wooden finishing, also in terms of improvement of the insulation and thermal comfort (*Photo* S. Pistidda, 2022)

Fig. 5.14 Some recently added elements are not coherent with the historic buildings, both in terms of materials used and architectural features. The new materials must be compatible with the existing ones, respecting the morphological and material character. The architectural elements should be designed to consider the harmonic relationship with the existing ones by avoiding copies of decorative elements (*Photos* F. Vigotti, 2022)

Fig. 5.15 A traditional wooden frame in a bad state of conservation and a replacement with PVC elements (*Photos* F. Vigotti, 2022)

5.2.6 Guidance for the Conservation of Materials Characterizing Historic Buildings (Basalt, Sandstone, Tuff, Wooden and Metal Elements)

The traditional materials of Gyumri (basalt, sandstone, tuff, wooden, and metal elements) need appropriate conservation interventions with the aim of stopping the degradation processes and prolonging life. Cleaning, consolidation and protecting works are allowed to reestablish consistency of the elements. The use of compatible materials in the interventions is a compulsory requirement to avoid the formation of other decay (Fig. 5.16).

No! Incompatible works on historical materials

Fig. 5.16 Grouting joints with incompatible materials (cement mortar) gives rise to efflorescence, detachment, and consequent degradation of the surfaces. Moreover, a grouting intervention like the one shown on the right decreases the tuff's natural texture and 3D perception (*Photos* S. Pistidda, 2022)

5.3 Technical Sheets for Maintenance, Conservation, and Reuse Works

This section illustrates the correct practices for the main conservation and maintenance interventions required by the buildings in Gyumri in relation to the general conservation status. Also, small interventions can profoundly transform and distort buildings if carried out without considering the historicity and characteristics of edifices. Some of them can be implemented directly by the inhabitants of the historic buildings; others necessarily require expert intervention (Fig. 5.17).

The *Technical sheets for maintenance, conservation and reuse works* present the main corrective interventions required for the conservation of the historic buildings in Gyumri. The sheets describe the most common problems, highlighting the damage to the buildings and giving possible solutions. It is suggested when the inhabitants can perform some interventions or when it is necessary to get professional assistance. The interventions are easily described with step-by-step instructions.

When performed without considering the peculiarities of historic buildings, small maintenance interventions can deeply transform the buildings, affecting their structures and causing collateral damage. The technical sheets give suggestions for correctly implementing the work to avoid such risks. The technical sheets guide inhabitants, artisans, or experts concerning small maintenance interventions that could support the conservation of the buildings.

The structure of each technical sheet is divided into two parts. The first section identifies the problem (decay phenomena or problems affecting the building). An easy and brief description is accompanied by pictures of the identified issues. The second section of the technical sheet provides a hands-on, facilitated, and concise operational guide to intervention that the inhabitants can perform. When self-maintenance is not sufficient and technical expertise is required, this is indicated in the technical sheet.

Works

7.3.1.1 Cracks grouting	■	■	■
7.3.1.2 Repairing deep cracks	■	■	
7.3.1.3 Consolidation with injections of hydraulic lime mortar	■	■	
7.3.1.4 Removal and replacement of a stone wall	■	■	
7.3.1.5 Punctual replacement of stones when strongly eroded or lacking		■	
7.3.1.6 Stone blocks joints grouting		■	■
7.3.2.1 Tuff façade cleaning with water and brush			■
7.3.2.2 Tuff façade cleaning with water and ammonium carbonate	■	■	
7.3.2.3 Removing areas of efflorescence		■	
7.3.2.4 Removing of biological colonization		■	
7.3.3.1 Treatment of metal elements (railings, canopies)		■	■
7.3.4.1 Dry cleaning and protection of wood surfaces		■	■
7.3.4.2 Treatments for insect attacks		■	■
7.3.4.3 Replace a missing wooden part		■	
7.3.5.1 Insertion of double frame in existing windows		■	■
7.3.5.2 Installation of new windows and doors		■	■
7.3.5.3 Indications for the insertion of new systems (gutters and drainpipes, antennas, air conditioners...)		■	■

Expertise

- ■ Inhabitants
- ■ Craftsmen
- ■ Experts (Architects, Engineers)

Fig. 5.17 A summary table of the available workforce in relation to the interventions described in the technical sheets (elaboration by authors, 2022)

The technical sheets could serve as a monitoring tool for self-maintenance interventions made by inhabitants and a guiding instrument for the building's anamnesis. Besides, the technical sheets can be used jointly with the section **General Suggestions** to double-check the quality of the interventions.

Technical sheets guide self-maintenance works. When needed, professional and technical expertise is clearly stated. In such cases, the assistance of skilled personnel is required to avoid damage and alterations to buildings' structure and appearance.

5.3.1 Masonry

5.3.1.1 Superficial Cracks Grouting

What is the problem?

There are various types of cracks. Superficial cracks do not represent a threat to the structural stability of the building but can cause damage to the facade surface, thus leading to loss of material. It is necessary to intervene by grouting the cracks.

Deep cracks on the masonry represent a lack of connection between the parts. They can also be caused by past seismic events. These discontinuities indicate a problem with the stability of the masonry and even the potential for structural collapse. Superficial and deep cracks also promote the entrance of water and other causes of decay, increasing the loss of cohesion and resistance in the masonry.

For this intervention, you need the technical assistance of a professional (structural engineer, architect) and a specialized craftsman.

What can you do?

If the crack is superficial (the width of the crack is narrow—hair cracking), you can proceed with grouting the cracks as follows:

- Protect the surface with masking tape.
- Gently clean the surrounding surface by using a soft dry brush to avoid any damage to the façade surface.
- Using a sprayer, wet the surface with distilled water (free from impurities).
- Using a spatula, fill the crack, leaving no empty spaces [hydraulic lime mortar, free of soluble salts and riddled sand (binder aggregate relation 1:3)]. Avoid using cement mortar.
- Do final grouting with natural hydraulic lime mortar and riddled sand (binder inert relation 1:2).
- Once the grouting is finished, clean the surrounding surface with a dry brush and pat with a damp sponge.

5.3.1.2 Repairing Deep Cracks

What is the problem?

Cracks are frequently diffused on the masonry surface and these mean lack of connection between the parts (Fig. 5.18).

These discontinuities indicate a problem with masonry stability and even the potential for structural collapse. They have to be checked by engineers and also monitored to understand whether the structural instability is in progress, or the cracks are the results of past damages.

They are also a concern because they promote the entrance of water and other causes of decay, increasing the loss of cohesion and resistance in the masonry.

Fig. 5.18 Cracks in stone masonry (*Photo* F. Vigotti, 2022)

What can you do?

- Protect the surface with masking tape.
- Gently clean the surrounding surface by using a soft dry brush to avoid possible damage to the surface of the façade.
- Using a sprayer, wet the surface with distilled water (free from impurities).
- Make deep injections of hydraulic lime mortar.

The mixture will fill the empty spaces within the masonry.

Before performing the injections, all cavities and minor fractures should be grouted to prevent the escape of the mixture.

Make holes in the wall with a drill and inject the mixture with a syringe. For deep injections, small plastic tubes may be inserted into the masonry. The injections should be made at a short distance to avoid overstressing the masonry.

The injections will be made up for rejection.

- Do final grouting with natural hydraulic lime mortar and riddled sand (binder inert relation 1:2).
- Once the grouting is finished, clean the surrounding surface with a dry brush and pat with a damp sponge.

Attention! To avoid imbalances, the injections should be made at low pressure, from top to bottom, from the outside to the center.

Dilute the mixture to facilitate penetration.

5.3.1.3 Removal and Replacement of a Stone Wall

What is the problem?

Sometimes, masonry is compromised by localized missing parts or large parts with damaged materials (Fig. 5.19).

When this happens, masonry loses its characteristics and mechanical strength.

Fig. 5.19 Missing parts in stone masonry (*Photo* F. Vigotti, 2022)

Different factors contribute to modifying its strength, leading to cracks and, in the worst cases, collapses.

Moreover, the lack of parts promotes other degradation processes.

What can you do?

The removal and replacement techniques are based on the removal of damaged elements and the substitution with other elements similar in shape, size, materials, and method of production.

In this way, it is possible to prevent the structure from collapsing.

First, it is important to:

- carefully evaluate the balance and stability of the masonry.
- make a correct selection of the replacement materials.
- carefully execute the replacement work to avoid cracks and collapses.
- Identify the parts to be removed. If needed, add provisional supports to prevent collapses or deformation.

Remove the deteriorated stones and mortars, taking care to proceed from top to bottom, working in small sections to avoid putting the stability at risk. The removal may be done with manual instruments or with the aid of simple tools such as hammers, drills, and levers.

Attention! Minimize stress on the structure to avoid further damage.

- Clean the cavity with a brush to remove rubble and dirt. It is also possible to use small instruments, such as scrapers and water.
- Choose new stones that have the same physical and mechanical characteristics as the existing ones.
- Provisionally place the new stones to check their shape and the result.
- It is possible to use small pieces of wood to facilitate the placement and removal.
- Wet the surfaces of the support and the stone with water.

- Place the new elements on a layer of hydraulic lime mortar (one part lime and three parts sand). Grout the joints between the elements with hydraulic lime compact mortar, then do a final clean of the surface.

Attention! When possible, reuse parts of the collapsed wall. The texture will be similar to the parts that have been removed.

- Avoid using cement and artificial mortar so as not to cause rejection of the existing masonry.

5.3.1.4 Punctual Replacement of Stones Strongly Eroded or Lacking

What is the problem?

The stones are often subjected to intense erosion, disruption, pulverization, and cracking due to different causes of degradation (age, pollution). The decay is manifested on the surface in different ways: separation of the joints, fractures, and erosion. The loss of parts can be accelerated until there is a complete lack of the elements (Fig. 5.20). Ultimately, the masonry becomes weaker, and this generally compromises stability.

The stone's integration and replacement reestablish continuity in the masonry to protect it from weathering.

What can you do?

It is possible to integrate the missing parts by applying natural lime mortar in successive layers or, in more severe cases, by totally replacing the elements.

- Remove the mortar joints around the damaged parts using a chisel and hammer until reaching the inner part of the wall.

Attention! During the removal work, be careful not to damage the adjacent areas.

Fig. 5.20 Lack of stone blocks (*Photo* F. Vigotti, 2022)

- Choose new pieces with the same physical and mechanical characteristics as the existing ones.
- Provisionally place the new pieces to check their shape and the result.
- It is possible to use small pieces of wood to facilitate the placement and removal.
- Wet the surface of the support and the replacement elements with water.
- Place the new elements on a layer of hydraulic lime mortar (one part lime and three parts sand). Grout the joints between the elements with hydraulic lime compact mortar and do a final cleaning of the surface.

5.3.1.5 Stone Blocks Joints Grouting

What is the problem?

The layers of mortar ensure cohesion between the stone blocks, giving stability to the masonry. In addition, the joints build a barrier against the penetration of water and other substances, extending the life of the materials.

In the worst cases, a deep erosion of the joints might cause structural problems to the building.

For this reason, it is important to integrate the missing or strongly eroded mortar joints (Fig. 5.21) by adding new material to restore their function as a connecting element [5].

You might need technical assistance to grout the joints if the erosion of the mortar is so advanced that the joints are not visible anymore (*missing joints*).

What can you do?

If the erosion of the joints is superficial, you can intervene by following these steps:

- Protect the ashlars with paper tape.
- Clean the joints with brush and water to permit contact with the new mortar. You can use a soft brush to gently clean the joints.
- Wet thoroughly to facilitate the setting of the mortar.
- Fill the joints by using a spatula, taking care to leave no empty spaces. Avoid using cement mortar and preferably use natural hydraulic mortar.

Fig. 5.21 Lack or strong erosion between the stone blocks (*Photo* F. Vigotti, 2022)

Once the grouting is finished, clean the surrounding surface with a dry brush and pat it with a damp sponge.

Attention! The joint must be identic to the existing one in dimension. The mortar must be inserted in the joint only and not leak over the stones.

5.3.2 *Facades*

5.3.2.1 Tuff Façade Cleaning with Water and Brush

What is the problem?

Every day facades are exposed to potentially damaging forces such as wind, rain, pollution, and humidity. Tuff is a porous material that can be easily subjected to phenomena such as rising dampness, condensation, and infiltration of water. Such phenomena might alter the appearance of tuff facades, resulting in efflorescence (presence of salts on surfaces), crust (a thick deposit), erosion, and pulverization (Fig. 5.22) of the stone blocks [6].

What can you do?

It is essential to regularly clean the facades to preserve the character of materials and prevent other kinds of decay. Small interventions carried out on a regular basis can preserve the facades over time.

When the façade has no decaying phenomena and shows deposits, you can clean it with brushes and water.

- Remove dirt using water (water barrel, preferably free of soluble salts, at aqueduct pressure) and brush with sorghum brushes in the meantime.
- The cleaning must be done from top to bottom for horizontal sections.

Attention! Before cleaning, it is necessary to protect the elements that could be damaged. Also, prepare a system for the collection and disposal of dirty water.

Fig. 5.22 Efflorescence and deposits on the façade surface (*Photo* F. Vigotti, 2022)

Fig. 5.23 Diffused and compact black deposits (*Photo* F. Vigotti, 2022)

5.3.2.2 Tuff Façade Cleaning with Water and Ammonium Carbonate

What is the problem?

A compact black deposit (Fig. 5.23) over many tuff facades in Gyumri is probably associated with biological colonization. Not only do these deposits jeopardize the polychrome of the tuffs, but they can also accelerate the deterioration of the stone material.

What can you do?

To clean the façade with water and chemical solutions:

- Prepare a solution made of water (free of soluble salts, possibly deionized), ammonium carbonate (10%), and quaternary ammonium (2%).
- Wear gloves and a mask and wipe the surface affected by the black deposit with sponges and cloth brushes soaked in the solution.
- Rinse the surface with water (free of soluble salts, possibly deionized).

5.3.2.3 Removing Areas of Efflorescence

What is the problem?

Efflorescence refers to the presence of salt crystals on the surface of masonry.

They appear as white powder or crystals, and they can be associated with other decay problems (Fig. 5.24). They are mostly related to the presence of water in masonry. They are transported from the inside to the outside until they crystallize on the surface. The crystallization of salts is destructive to the materials, causing erosion and detachment on the surface. The soluble salts may be of different types, causing different problems.

What can you do?

It is essential to remove areas of efflorescence to maintain the condition of the masonry.

Fig. 5.24 Diffused efflorescence (*Photo* F. Vigotti, 2022)

In addition to cleaning the surface areas, you must also determine why water is present and address the cause.

Before cleaning, it is essential to identify and remove the cause of the water problem to prevent salts from re-forming.

To find the source of the water:

- Check for discharging stormwater.
- Check the waterproofing of the roof.
- Check for water leakage in the interior (possibly due to system conditions).
- Check for rising dampness.

Clean the surface by dry brushing. The brushing should be done from top to bottom.

- Rinse the surface with water (free of salts and possibly deionized).
- Allow the surface to dry. If the salts reappear when dry, repeat brushing and washing until they completely disappear.

In more severe cases:

- apply packs of absorbent paper pulp: prepare the pack by soaking facial tissue in distilled water until it has a pasty consistency. This will make it easier to apply it.
- leave on the surface until dry.
- after the application, do a final washing with water to remove all the material.

5.3.2.4 Removing Biological Colonization

What is the problem?

In a humid environment, algae, mosses, and lichens can grow on the surface of materials: from the surface, they can also enter the pores of plaster and bricks, changing the color of the wall and contributing to its degradation by causing disaggregation, decohesion and other kinds of decay (Fig. 5.25). These biological species proliferate mostly in areas exposed to weather effects and subject to rising damp or especially

Fig. 5.25 Biological colonization (*Photo* F. Vigotti, 2022)

humid, such as below gargoyles, on the north-facing facade and within the humid environment of cracks and cavities.

What can you do?

It is important to remove microorganisms to save the condition of the masonry.

In addition to cleaning the surface areas, you must also determine why water or humidity is present and address the cause.

- Wear gloves, glasses, and a mask.
- Clean the surface by spraying a biocide (bleach is also good).

Attention! A good biocide must have the following characteristics:

be colorless or transparent, have a low degree of toxicity, be degradable over time, not able to cause physical or chemical actions on the wall structure, and not able to leave residues after application.

- Let it work for the necessary time (depending on the product used).
- Soften the brush with water, then brush from top to bottom.
- Choose the hardness of the brush according to the surface.
- After the application, do a final washing with water to remove all products.

5.3.3 Iron Works

5.3.3.1 Treatment of Metal Elements (Railings, Canopies)

What is the problem?

Metal elements, such as railings and canopies, are frequently used in the entrances of the buildings set in Kumayri Historic Reserve. Canopies are put in place to shelter inhabitants from adverse weather conditions while accessing their houses; railings are mostly used as safety and protection measures in balconies and stairways. After

Fig. 5.26 Atmospheric corrosion of a metal gate (*Photo* S. Pistidda, 2022)

prolonged exposure to rain, wind, and other endangering factors, rust might affect these elements (Fig. 5.26).

What can you do?

General cleaning of the elements is recommended on a regular basis to get rid of deposits. Regular maintenance of metal elements could be performed quickly by brushing specific products directly on railings and canopies to prevent rust formation.

Antirust treatment

- Clean the rust layer on the iron using a dry brush.
- Treat the clean surface by applying citric or phosphoric acid with a rough plastic brush.
- Wash the surface with water.
- Dry with a cotton cloth.
- Use a brush to apply a layer of rust converter available on the local market.
- Repeat this step several times, with two or more thin layers.
- After the antirust treatment, apply colored paint if needed.

Replacement of the fittings

- Remove the iron fittings taking care not to damage the surrounding parts.
- Replace them with elements compatible with the existing ones in dimension and form.

Check and repair the attachment points on the wall

- After replacing the elements, check and repair the attachment points, repointing them with lime mortar to avoid water infiltration.

5.3.4 *Wooden Works*

5.3.4.1 Dry Cleaning and Protection of Wooden Surfaces

What is the problem?

Wooden elements are frequent in most historic buildings set in Kumayri Reserve. Often, these elements are decorated by skilled craftsmen and represent a distinctive trait of the appearance of historic buildings.

However, in time, wooden elements might be subjected to several decay phenomena: deposits, insect attacks, color retrieval, fissures, and breaking of some parts (Fig. 5.27). Regular maintenance is needed to preserve these elements.

What can you do?

When the surface has patina and deposits, it is essential to clean the elements without damaging the wood [4]. After cleaning, it is also possible to use some protective coating on the wooden surfaces.

- Dry clean by dusting with nylon or sorghum brushes. The cleaning must remove the deposits of atmospheric particles, pollution dust, and other extraneous residues.

  Attention! Clean from top to bottom for horizontal fields.

- Brush following the direction of the wood grain to avoid scratching the surface.
- Brush or spray two layers of synthetic, transparent, and colorless coating on all wood surfaces.
- Apply a UV-resistant coating on external surfaces, using a compatible tint if the wood element was colored.

Fig. 5.27 Deposits, color retrieval, and fissures in the wooden gate (*Photo* F. Vigotti, 2022)

Attention! Linseed oil might also be used as a protective coating, but that will require more maintenance in the future.

5.3.4.2 Treatments for Insect Attacks

What is the problem?

Insects can attack some parts of wooden elements.

Wood decay caused by insects can lead to progressive loss of mass, resistance, and hardness, and changes in color and appearance.

What can you do?

First, it is important to assess how severe the decay is.

Sometimes the wooden elements attacked by insects only need cleaning and treatment; however, if the decay is profound, it may be necessary to strengthen or replace all or a portion of them.

Make a preliminary diagnosis to verify the extent of decay due to insects.

If the problem is not profound:

- Dry clean by dusting with nylon or sorghum brushes, vacuum cleaning, and compressed air on the near surface. The cleaning must remove the deposits of atmospheric particles, pollution dust, and other extraneous residues.
- Treat with synthetic products or oil-based natural products [7].

Attention! Using natural products will require more maintenance in the future.

If the decay has progressed to loss of mass, resistance, or hardness:

- Strengthen the wood with synthetic products or animal glue.
- Replace a damaged wooden part with a new element having the same dimensions and the same wood type as the original parts.

Attention! Using the same wood type allows to have good compatibility with the rest of the features.

5.3.5 Reuse Works

5.3.5.1 Insertion of Double Frame in Existing Windows

What is the problem?

Many historic buildings in Kumayri Historic Reserve keep old wooden window frames (Fig. 5.28). Unexpectedly, sometimes a double window is used: one is flush with the facade, and the other one is more internal. This trick allowed the creation of an air chamber and improved thermal insulation. This technology should be preserved, as it provides better insulation than any other contemporary window. Further to the

Fig. 5.28 Window with wood surface issues and single-pane glass (*Photo* S. Pistidda, 2022)

properties of this system, it is possible to insert a second pane of glass in the same window.

To perform this intervention, you will need the technical assistance of trained artisans.

What can you do?

Restoring the existing wooden frames is always the priority. To improve thermal resistance, it is possible to proceed in two ways, depending on the starting conditions.

If there is a double window (one external and one internal), it is only possible to restore the existing wooden frames because the thermal conditions are assured by the double system.

If there is no internal double window, it is possible to restore the existing frames by inserting a double glass. Given the delicacy of the intervention, a skilled artisan must perform the work.

When the internal conditions allow it, it is also possible to build a second window from scratch.

Fig. 5.29 Ancient windows preserved until now (*Photo* F. Vigotti, 2022)

5.3.5.2 Installation of New Windows and Doors

What is the problem?

Many historic buildings in Kumayri Historic Reserve keep old wooden doors and window frames (Fig. 5.29). If these elements are too damaged to be preserved, replacing them with new elements that can grant thermic comfort, and safety is necessary [12].

What can you do?

When installing the new elements, choosing the material and the form of window frames and doors is important. The new insertions must be made of the appropriate material and color to be compatible with the existing building. You will need the technical assistance of a skilled craftsman to perform this intervention correctly.

In terms of architectural shape, the new frames can draw inspiration from the existing ones by respecting the elements' partitions, rhythm, and dimensions.

An abacus of the windows and doors of all the historic houses of Gyumri can be useful to preserve the characters of these elements.

Attention! Every intervention must be preceded by a careful study of the existing characters!

5.3.5.3 Indications for Inserting New Systems (Gutters and Drainpipes, Antennas, Air Conditioners, etc.)

What is the problem?

Historic buildings might not be equipped with contemporary systems such as antennas, satellite dishes, and air conditioners (Fig. 5.30). It is possible to install

Fig. 5.30 Satellite dished and electrical cables on the main facade (*Photos* F. Vigotti, 2022)

these elements in historic edifices, but some attention is required when choosing their positioning over facades and roofs [15].

The same attention is required when it is necessary to replace existing gutters and drainpipes when these are too damaged to be preserved. The new elements must be in an appropriate material and color to be compatible with the existing building. You might need the technical assistance of a skilled craftsman to perform this intervention correctly.

Attention! A proper drainage system is essential to protect the masonry from water damage.

What can you do?

Antennas, satellite dishes, and air conditioner plants should NOT be placed on the main façade. If possible, place these elements on the roof of your house; if it is not possible, choose a position that does not impact the appearance and the structure of the main façade (e.g., place the elements on the courtyard-facing façade). Do not repeatedly pierce the façade to accommodate the new elements. Instead, put the new elements close to each other, asking for the help of a technician. Ask for help installing the elements supports in the masonry to avoid damaging the facades. Collect the electric wires in a pipe to prevent cables from running freely (possible danger).

5.4 Self-monitoring Checklist for the Inhabitants

This tool is addressed to the inhabitants to develop an awareness of the historical heritage in which they live [4].

By answering simple questions, they can check whether their houses are historic, safe (from a structural point of view), and have elements to preserve.

The checklists can also be used as a questionnaire for interviews or as an instrument to develop participatory processes.

Do you live in a historic building? (end of XIX century—1960s)

o Yes
o No

Is your house made of tuff?

o Yes
o No

Do you see cracks?

o Yes
o No

If you see cracks like these, call an expert!

Does your house have a masonry structure (tuff)?

- Verify that the mortar joints are in good condition and repair them when they are eroded/missing (see technical sheet Stone blocks joints grouting). Do this with the help of a skilled technician. If the joints are eroded or missing, this may cause instability in the entire building.
- Check if the tuff blocks are in good condition. If some blocks are deeply eroded, proceed to replace deteriorated elements. Do this with the help of a skilled technician (see technical sheet Removal and replacement of a stone wall and Punctual replacement of stones when strongly eroded or lacking).

Is your building subjected to partial collapse?

Ask an expert to evaluate removing the collapsed portion of the structure and its correct stabilization to avoid further damage to the building.

Does your building have metal elements (canopies, gutters, railings)?

- Check the metal elements for the presence of rust or deposits. If so, proceed as explained in the technical sheet. Treatment of metal elements (railings, canopies).
- If some parts are widely corroded, you will need to replace them. In case of damaged railings, the missing part can be replaced carefully, using a compatible element (same form and dimension as existing ones).

Do you want to improve the living conditions of your home?

Do you want to improve the internal climate conditions of your house?

Check the condition of doors and windows: if they are still in good condition, but you can experience air leaking, consider the insertion of a double frame in existing windows, as explained in the technical sheet.

If doors and windows are damaged, consider substituting these elements with new, more performing but compatible ones, as expressed in the technical sheet Installation of new windows and doors.

Do you want to adjust your home's gas or electrical system?

Do you want to install an air-conditioning system?

Check the most rational system to gather and install gas and electrical systems. Prevent wires and pipes from running freely on your building façade; instead, group them in a dedicated duct, as explained in the technical sheet. Indications for the insertion of new systems (gutters and drainpipes, antennas, air conditioners).

Do you want to avoid water infiltration into walls?

Check the condition of the roofing and walls.

If cracks are present, water infiltration will likely leak inside the building. At the same time, it is important to maintain an efficient system of drainage (gutters and drainpipes). Follow the indications as provided in technical sheet "Cracks grouting".

References

1. Canada's Historic Places (2010) Standards and guidelines for conservation of historic places in Canada. Her Majesty the Queen in Right of Canada
2. Department of Arts, Heritage and the Gaeltacht (2011) Architectural heritage protection. Guidelines for planning authorities. Government of Ireland, Dublin
3. Directorate General Central Public Works Department (2013) Conservation of heritage buildings—a guide central public works. Department New Delhi
4. Giambruno M, Pistidda S (2015) The walled city of Multan, Guidelines for maintenance, conservation and reuse work. AltrAlinea, Firenze
5. Historic England (2012) Practical building conservation: mortars renders and plasters. Routledge, London
6. Historic England (2012) Practical building conservation: stone. Routledge, London
7. Historic England (2012) Practical building conservation: timber. Routledge, London
8. Historic England (2024) Practical building conservation: conservation basics. Routledge, London
9. ICOMOS Singapore (2017) Introduction. In: Conservation technical handbook, vol I. URA, Singapore
10. ICOMOS Singapore (2018) Facades. In: Conservation technical handbook, vol III. URA, Singapore
11. ICOMOS Singapore (2019) Structure. In: Conservation technical handbook, vol IV. URA, Singapore

12. ICOMOS Singapore (2019) Doors and windows. In: Conservation technical handbook, vol V. URA, Singapore
13. ICOMOS Singapore (2020) Paints and coatings. In: Conservation technical handbook, vol VI. URA, Singapore
14. Marshall J (1923) Conservation Manual, reprinted by Asian Education Service. New Delhi (1990)
15. U.S. Department of the Interior National Park Service, Heritage Preservation Services (1997) The secretary of the interior's standards for rehabilitation and illustrated guidelines for rehabilitating historic buildings. Washington DC

Chapter 6
Starting from Gyumri. Methods and Tools for Planning Historic Centers

Abstract This chapter reports on two of the studies conducted for the city of Gyumri, which can serve as a basis for setting up preliminary analysis tools for drafting Plans for historic centers in other contexts as well. Appropriately adapted to the specificities of the place, they provide an essential reference framework for comprehensive knowledge of the urban fabric, the buildings, and the open spaces that compose it. The first tool consists of an illustrated directory of primary problems of decay. This photographic directory, organized by problems and annotated with long captions, has a dual utility. On one hand, the simple language that accompanies the images makes it understandable even to the residents; on the other hand, it allows the municipal administration to quickly verify over time whether the problems are worsening or being resolved. The second tool contains the guidelines for the minimum quality that base maps should have to conduct studies on a historic center, as well as the description of some of them necessary to tackle the drafting of the Plan with sufficient knowledge of the place. The two tools, not particularly innovative for contexts where there is a long-established practice in drafting specific plans for "historic centers", can serve as a methodological guide in the so-called emerging countries where, with few exceptions, normally the only instrument is typically a master plan that does not differentiate tools and regulations for the new parts and the older parts of the cities.

Keywords Methodological tools · Illustrated glossary of decay · Planning historic centers

6.1 Proposed Studies for the Inventory and Assessment of the Historic Core

Implementing a proper planning process for a historic city needs to start with the deep study of appropriate cartography that can represent the characteristics and specificities of the context in a correct way [1, 2]. This part illustrates the proposal cartography, together with some in-depth studies on the conditions of the historic center, which could help represent it in its entirety and its characteristics.

M. Giambruno et al., *Heritage-Based Tools for the Sustainable Urban Development of Historic Centers*, PoliMI SpringerBriefs, https://doi.org/10.1007/978-3-031-75259-9_6

Appropriate graphic scales are suggested to represent the elements of the historic center. The cartography is required to draft subsequent data-gathering detailed studies described below.

Basic cartography and in-depth studies are organized according to a schematic sheet containing the following information:

What: the goals of the assessment.
Why: the description and the usefulness of the map, the purposes of the specific assessment.
How: the survey techniques and the proper scales of representation.
Relationship: other studies with a connection to the phase.
Warnings: possible limits and potentialities of the representation and critical issues in the execution.

The description is complemented with illustrated images that exemplify the type of graphic representation of the cartography and the information that each map may contain concerning the drawing scale.

6.1.1 Base Cartography and Maps

6.1.1.1 Territorial Scale Maps

WHAT

This phase aims to organize, read, and interpret the territory's data assets.

The map is oriented to identify the territorial-environmental system: the perimeter of the Municipality, the borders of the historic center, the specific elements of historical interest, the green areas (both private and public), the environmental characters, the areas of recent expansion, the infrastructural elements (railway, main streets, secondary streets, etc.).

WHY?

The territorial scale analysis aims to represent the data of territorial and environmental systems in a synthetic form to understand the relations among the built environment, historic elements, infrastructures, and landscape.

HOW

The analysis can rely on the most recent technical cartography on scales 1:2000 (Fig. 6.1) and 1:1000 (Fig. 6.2) for different details of the information. An aerial photogrammetric map, a roofing map, or a cadastral map with the silhouettes of the buildings can be used as a cartographic basis. The scale 1:2000 is helpful in representing general information about the territory and the environment, such as the geological and morphological aspects, the system of territorial axes, and the main infrastructures.

Fig. 6.1 The map shows the municipal boundary, historic centers, and characteristics. A similar map is helpful to illustrate the municipal territory in its entirety and understand its consistency and the relationship between new and old parts. Map of the territorial-environmental system, Reconstruction plan for the municipality of Fontecchio, Italy (© M. Boriani, M. Giambruno, R. Simonelli . R. M. Rombolà, F. Maroldi)

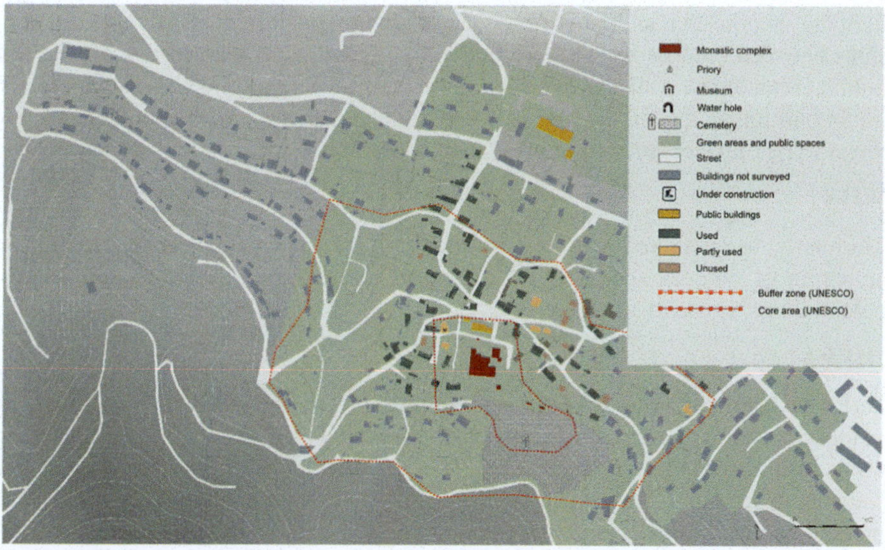

Fig. 6.2 The map (scale 1:1000) refers to the functional zoning, properties, relevant historic elements, and landscape elements. Guidelines for urban development improvements of the potentialities, Haghpat, Armenia, Masterplan for the World Heritage Sites (elaboration by authors)

The scale 1:1000 is aimed at the second level of investigation and used to identify punctual elements of the urban system (monuments, aspects of historical interest).

The data is returned by thematic maps where the different aspects of the territory are visually summarized.

RELATIONSHIP

With the identification of multiple data, the maps can be overlapped to better understand the elements of the urban system and their correlations.

WARNINGS

The map scales must be chosen according to the detail of the information to be represented. These initial maps constitute the preparatory studies for subsequent analyses; therefore, they must be designed according to the specific characters of the context, avoiding standardized representations but immediately identifying the elements of the system.

6.1.1.2 Detailed Scale Maps

WHAT

The detailed maps represent the second level of in-depth analysis, oriented to highlight the specific characteristics of the urban texture: the historic fabric, the relationships between outdoor spaces and the built environment (urban fronts, private and public green areas, walls, streets, squares), the historic buildings (monuments and minor buildings), the building uses, the access and path system, relevant perspective views, type of pavements, lighting system, services.

WHY?

The survey is conducted at the urban scale and allows the urban system to be read as a sum of outdoor spaces and built environments to understand relationships, potentialities, and critical elements.

HOW

Different thematic maps (historical evolution, service network, characters of the outdoor spaces, differences in levels, paving, covered passages, lighting system, and urban furniture, buildings use, characters of the fronts, materials, historic elements, state of conservation of the facades, building systems equipment, etc.) are produced based on a map of the ground floors—or, in its absence, of the roofing (Fig. 6.3).

RELATIONSHIP

The different thematic maps can be overlapped to cross information.

Fig. 6.3 Maps of the historic centers, Nesolio (left) and Fontecchio (right), Italy. The drawing of the entire historic center at a proper scale with the identification of the ground floors allows for the critical analysis of the urban system through a joint reading of the outdoor spaces and the urban fabrics. In particular, the map identifies the building units (elaboration by authors)

WARNINGS

The work can be realized with the support of the GIS system to combine quantitative and qualitative data.

6.1.1.3 Survey of Building Facades

WHAT

The survey of the building's facades aims to understand the characters and the conservation state of the fronts at the urban scale to understand connections between buildings, dimensional relationships, skyline, relations between historic buildings, and new insertions, materials, and decay phenomena.

WHY?

The urban fronts contribute to defining the quality and character of streets, squares, and all the elements of the urban texture. Representing entire urban fronts assists in providing valuable data to reflect on the urban system. This basis helps to implement conservation strategies on the existing built heritage and to guide both its maintenance and the new insertions (for example, to plan the colors of buildings and urban furniture in an integrated and systemic vision).

HOW

In addition to the traditional survey techniques, orthophotos can be obtained via photogrammetric techniques (such as Agisoft Metashape Pro) or laser scanners (Fig. 6.4). Some recent smartphones have been equipped with a Lidar sensor. The

Fig. 6.4 Urban fronts of a street realized with Laser Scanner technique. Sarafa Bazar regeneration, Sustainable Social Economic and Environmental revitalization in the historic core of Multan City, Pakistan (© M. Boriani, M. Giambruno, C. Manfredi, R. Mastropirro, R. Gabaglio, S. Pistidda, M. Tasinato)

proper scale for the restitution of entire urban fronts is 1:100 or 1:200, and it is oriented to represent complete urban streets to highlight the relationships among the buildings.

RELATIONSHIP

The orthophotos represent the basis for other investigations: crack pattern survey, material decay survey, conservation projects, architectural projects at the urban scale, and maintenance instructions (colors, materials for repair, and replacement).

WARNINGS

When made with the appropriate techniques, an orthophoto serves as an accurate and measurable representation of reality. If realized with other empirical methods (such as photo rectification with Photoshop), attention should be paid to distortion phenomena.

6.1.1.4 Three-Dimensional Representations and LIDAR

WHAT

The LIDAR technology (light detection and ranging) allows high-resolution three-dimensional images, starting from a digitally processed point cloud. The technique helps to identify the morphology of the territory (differences in levels and environment) and the characters of the buildings. The three-dimensional model (Fig. 6.5) guarantees object measurability and can be digitally processed for many applications (profiles, sections, elevations, 3D views).

WHY?

The three-dimensional representation is beneficial for analyzing complex systems, as it allows to simultaneously manage a lot of information. Moreover, together with the survey of the current state, it will enable the development of a monitoring system.

Fig. 6.5 Three-dimensional reconstruction of the village of Cornello dei Tasso (Bergamo, Italy) (© Courtesy by F. Fassi, 3DSurvey Group, Polimi)

HOW

The three-dimensional representation can be obtained either from photogrammetric maps or from laser data. Once the scans are produced, orthophotos can be extracted in the post-processing phase according to different plans.

RELATIONSHIP

Once properly processed, the points cloud can serve as the basis for a Spatial Information System, similar to a GIS, supporting all the investigations.

WARNINGS

The data acquisition and processing must be carried out considering the survey's objectives with the specific characteristics of each technique.

6.1.1.5 G.I.S. of the Historic Center

WHAT

Historic centers are complex systems in continuous evolution and transformation. For this reason, their knowledge and management require open and flexible tools that are able to manage different information and intertwine them. The G.I.S. systems (Geographic Information System) allow the representation of different levels of knowledge of the territory (Fig. 6.6) and overlap the data by creating multiple thematic readings (built environment and historical, natural, and environmental characteristics).

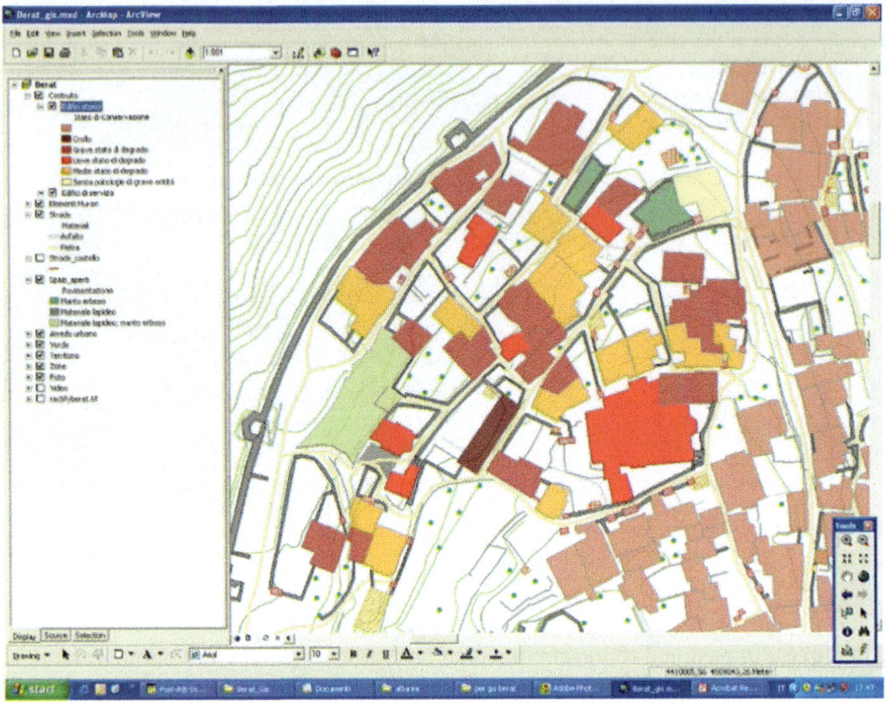

Fig. 6.6 G.I.S. of the historic center of Berat (Albania) (© M. Giambruno, R. Simonelli, M. Palo)

WHY?

The G.I.S. systems allow for the simultaneous management of information, both vectorial and quantitative, while continuously updating them over time (Fig. 6.7). Starting from data entry, the G.I.S. allows for infinite combinations of attributes. Only in-depth knowledge can guide the involved subjects to protect the existing contexts and to manage transformations consciously. As for the possibility of controlling the complexity, the G.I.S. can represent good support in decision-making processes.

HOW

The G.I.S. tools are conceived to store, process, analyze, manage, and represent geographic data. The system can arrange a vast amount of information (collected in a database) to precise and geo-referenced coordinates of the territory. Thanks to the possibility of connecting and crossing all this information, the G.I.S. can create new data to better understand the existing conditions of the context. For the construction of the G.I.S., the cartography must be geo-referenced.

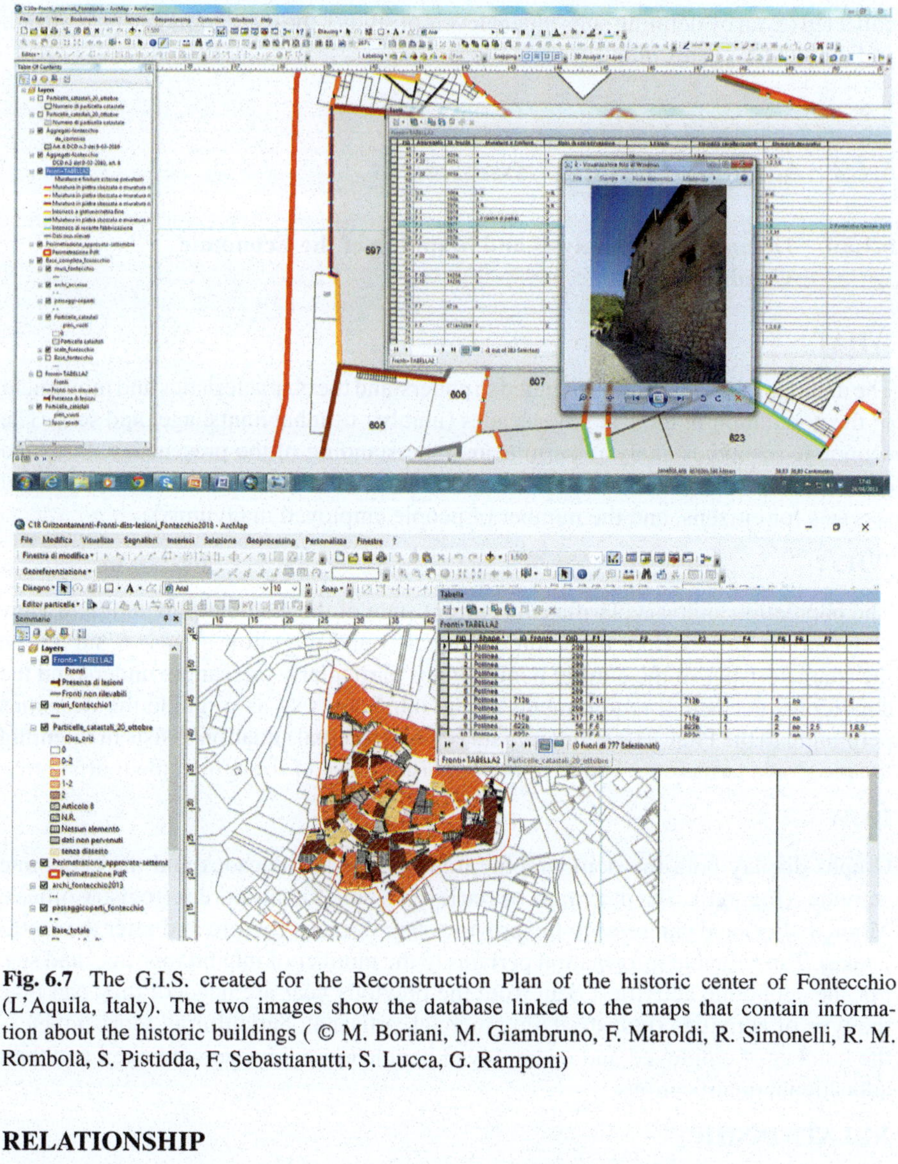

Fig. 6.7 The G.I.S. created for the Reconstruction Plan of the historic center of Fontecchio (L'Aquila, Italy). The two images show the database linked to the maps that contain information about the historic buildings (© M. Boriani, M. Giambruno, F. Maroldi, R. Simonelli, R. M. Rombolà, S. Pistidda, F. Sebastianutti, S. Lucca, G. Ramponi)

RELATIONSHIP

The evolution of the G.I.S. system includes the WebGIS (tool extension allowing the sharing of information on the web), the 3d GIS (real modeling of urban scenarios, involving a consequent increasing complexity to be managed), and virtual reality.

WARNINGS

The G.I.S. system is still a tool and must be carefully designed to achieve specific objectives. In the preliminary phase, it is thus essential to clarify the data to be collected (even by defining subsequent levels of study), the overlapping to be realized, and the specific goals of every thematic reading.

6.1.2 Additional Detailed Assessments

6.1.2.1 Demographic Analysis and Mapping of the Economic Conditions

WHAT

The demographic analysis is intended to understand the characteristics and movement of the population in the last few decades (number of inhabitants, age, and sex). The economic resources analysis permits an understanding of the presence, nature, and quality of the financial activities in the city (local and artisanal activities, essential services for citizens, and the number of people employed in Gyumri).

WHY?

The population analysis clarifies the current state of the resident population in view of important events (like the 1988 earthquake) and migration flows. A survey on the characteristics of the current population is particularly relevant to understand the conditions, needs and requirements of the inhabitants so as to guide the decisions pertaining to the Plan. The economic survey of the actual situation assists in orienting the proposals to increase employment opportunities and strengthen the tourist offer.

HOW

Graphs display both the demographic trend of the last decades and the economic activities (Fig. 6.8). The data must be retrieved from official sources (registry offices, Municipality) and can refer to the previous five decades (before and after the earthquake). The valuable information pertains to the number of inhabitants, age, and sex. For the economic activities, data from the previous year are needed; they can map the type of activities (industrial, commercial, artisanal, agricultural, touristic, etc.), the number of employed and unemployed people (sorted by sex and age), and their educational qualifications.

RELATIONSHIP

The obtained data can be interlinked with maps and graphs to realize cross-readings and themes.

WARNINGS

The data can present different levels of depth and accuracy. Therefore, it is necessary to check all the sources and the status of available information to compare similar data and to decide the type of graphs and tables.

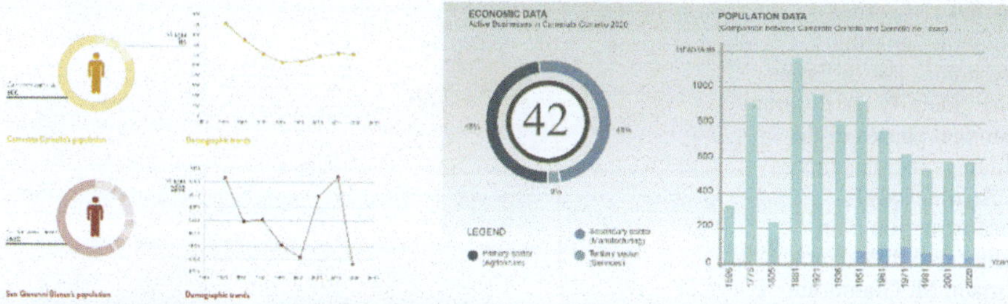

Fig. 6.8 Demographic trends and economic data, graphs, and schemes. Cornello dei Tasso, Bergamo, Course of Architecture Preservation Studio, Politecnico di Milano, prof. S. Pistidda, F. Fassi, M. Bruggi (© Survey by C. Bessega, G. Buratti, B. Marca, M. Mingardi, D. Topkurulu at left and B. Zuccarelli, C. Nessi. G. Kocak, K. C. Perino, Z. Kocabas at right)

6.1.2.2 Construction Phases of the City, Buildings, and Outdoor Spaces

WHAT

The historical analysis represents the first essential step in understanding a city (Fig. 6.9). The research investigates the entire life of the city, from its origin to the present day, collecting data related to its transformation to the physical, functional, and distribution changes. Reading the historical context in which the city was built is a fundamental step to acquiring knowledge of its buildings and outdoor spaces. The urban context is not a static monument but is subject to dynamic forces in the economic, social, and cultural spheres that shaped it and keep shaping it. Studying the transformations of the entire human environment with its tangible and intangible qualities permits us to balance the ability to preserve and innovate the city, subordinating these changes to the conservation of its specific identity.

WHY?

Analyzing the construction phases of the buildings (Fig. 6.10) and outdoor spaces of a city is useful to understand the contemporary city that appears to us as a complex system, which is the result of a continuous stratification process where the different ways of living have left indelible signs on the territory.

The urban fabric looks like a text being continuously written and overwritten and containing the traces of transformation processes. The stratification process leaves signs on the city and acknowledging them is the first level to approach complexity. This lecture allows us to transform this process into a project respectful of the natural vocation to transformation. The goal is also to trace what the "possible margin of transformation" is.

HOW

This analysis uses indirect historical sources such as bibliography, archival and iconography analysis, and the reading and comparison of historical city maps or old aerial photographs of the territory. The construction of historical thematic maps with

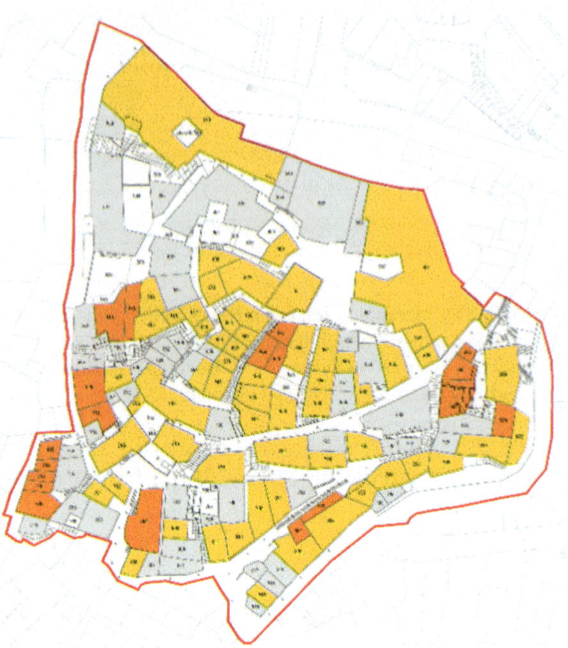

Fig. 6.9 Historical evolution: building thresholds. The different colors identify the different historical phases of the buildings. Fontecchio (L'Aquila, Italy), Reconstruction Plan, Politecnico di Milano (© M. Boriani, M. Giambruno, F. Maroldi, R. Simonelli, R. M. Rombolà, S. Pistidda, F. Sebastianutti, S. Lucca, G. Ramponi)

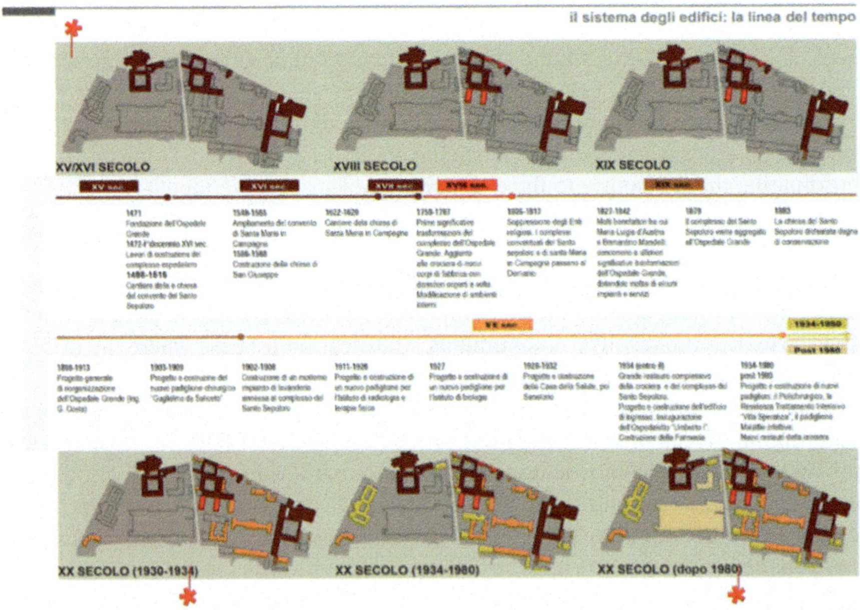

Fig. 6.10 The system of buildings: timeline. The grey base represents the current cartography, and the different colors of the buildings show their historical phases from the XV–XVI century to the XX century (elaboration by authors)

different phases of the urban settlement evolution is crucial to represent the placement of historical facts in relation to the space. The current cartography must be compared with the historical one (according to the main identified historical phases). In this way, it is possible to identify in which historical cartography a single building, street, or open space appears for the first time and to see its evolution until the current situation. Following this research, the different historical phases of the buildings can be represented with different colors based on current cartography. This reading can communicate the changes and territorial transformations that have taken place over the years. The data must be retrieved from official sources and can be useful to write a historical register that explains the most important phases of the city's transformation. In some specific situations, a direct survey to identify any urban stratifications (wall systems, etc.) with some archaeological investigations proves useful.

RELATIONSHIP

The obtained data can be interlinked with maps to realize cross-readings and thematic maps.

WARNINGS

It is advisable to investigate the city's history first so as to identify the main historical phases that have led to significant changes in the city's layout.

6.1.2.3 Underground Utilities and Lighting Systems for Outdoor Spaces

WHAT

The study of underground utilities and lighting systems for outdoor spaces is necessary to understand how essential and indispensable infrastructures are for the city's areas. It is crucial to know how the distribution of these elements is managed and whether the infrastructure network (sewers, electricity, water, gas, internet, etc.) reaches every part of the city, guaranteeing an adequate quality of life for all inhabitants.

WHY?

The analysis of networks, underground utilities, and lighting systems for outdoor public spaces gives essential indications of the quality of public and private spaces. For example, adequate lighting of outdoor public spaces significantly affects their use, safety and architectural quality. Furthermore, the study of these infrastructures is significant to understanding how modern a city is and how close it is to the principles of energy and economic sustainability.

HOW

The analysis is meant to investigate whether there are thematic maps in the municipality to know the distribution network of sewers, electricity, water, gas and internet (Fig. 6.11). Furthermore, it is possible to proceed with surveys to understand the distribution networks of these infrastructures and the localization of their single components. As it happens in many parts of Gyumri, the external distribution systems can majorly impact the outdoor public spaces; therefore, it is crucial to analyze their characteristics and quality. Thus, the mapping activity must consider the distribution system, any control units, the branch system to the single housing units and the single system components. As for electricity, for example, it is important to identify the mapping of the network of underground wires or exteriors, meters, and individual lighting bodies in public spaces. The survey also must determine whether a street lighting system exists and, if so, whether it works and is sufficient.

RELATIONSHIP

The obtained data can be interlinked with maps to realize cross-readings and thematic maps and can be useful for planning adequate maintenance of the different infrastructures.

WARNINGS

Many buildings can have obsolete equipment; therefore, it is essential to also map the quality and decay of the infrastructures.

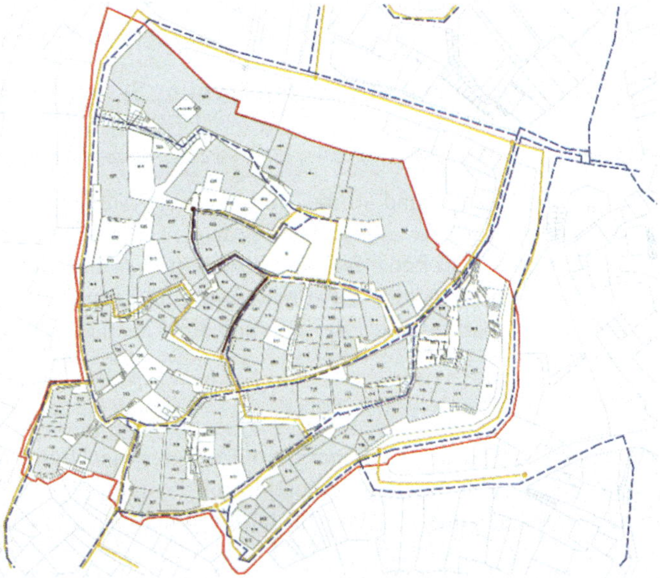

Fig. 6.11 Networks, underground utilities. The yellow line identifies the gas system, the blue line the water system. Fontecchio (L'Aquila, Italy), Reconstruction Plan, Politecnico di Milano (© M. Boriani, M. Giambruno, F. Maroldi, R. Simonelli, R. M. Rombolà, S. Pistidda, F. Sebastianutti, S. Lucca, G. Ramponi)

6.1.2.4 Analysis of Mobility and Pedestrian Flows

WHAT

The mobility and pedestrian flow analysis is important to understand both the distribution and balance between driveways and pedestrian streets and how the city is served by public transport and its accessibility.

These flows affect the quality of life as their good management can guarantee less car traffic, less pollution, and an improvement in the quality of public spaces.

WHY?

The mobility and pedestrian flow systems significantly affect the design and quality of public spaces. Studying these flows helps to understand the distribution of driveways, pedestrian areas, cycle paths, and interchanges with public transport vehicles, guarantee good accessibility to all parts of the city, and ensure livable public spaces. The management of a city's mobility impacts the possible acquisition of environmental sustainability values and characterizes important elements relating to the design of public spaces.

HOW

The analysis moves from building adequate maps in which the different dimensions of the street, the traffic routes, the driveways, the pedestrian areas, the parking areas, the distribution of public transport, and their stops can be recognized (Fig. 6.12). This is useful to know the current state of the mobility and pedestrian flow system.

The data can be collected from existing sources (existing official maps, online maps, official city plans, etc.) or verified/collected through surveys.

RELATIONSHIP

The obtained data can be interlinked with other maps to realize cross-readings and thematic maps. For example, understanding the link between the pedestrian street system and the green areas can prove useful.

WARNINGS

The mobility plan must be processed on a big scale to understand the criticalities and needs of the city.

6.1.2.5 System of Property Ownership

WHAT

Buildings and outdoor spaces in a city can have different ownership regimes.

Properties can be divided into three main categories: public (i.e., owned by the State and its reference bodies), semi-public (i.e., owned in part by the State and in part by private citizens), and private (i.e., owned by individual citizens, businesses, companies, and religious bodies).

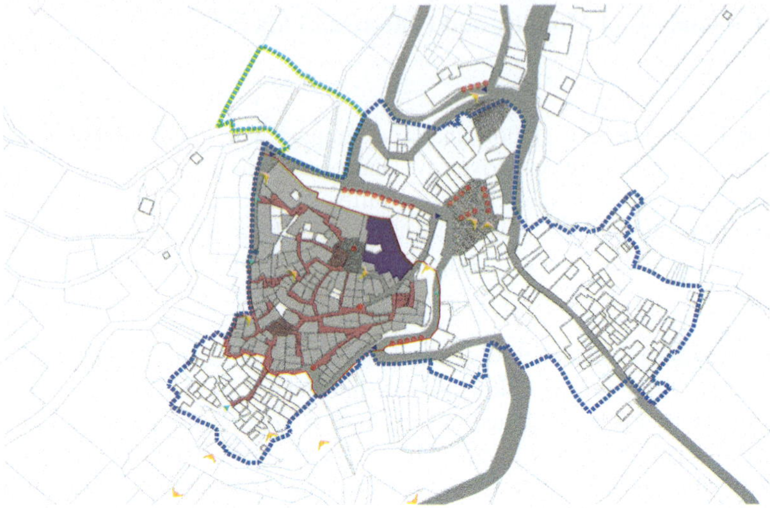

Fig. 6.12 Character and potential of the outdoor spaces. The map shows the viability (driveways in grey, pedestrian streets in red), the parking lots (the red spots represent the regulated ones, and the blue spots represent the illegal ones), and the accesses to the historic center (the green arrow represents the pedestrian accesses and the blue ones the driveways). Fontecchio (L'Aquila, Italy), Reconstruction Plan, Politecnico di Milano (© M. Boriani, M. Giambruno, F. Maroldi, R. Simonelli, R. M. Rombolà, S. Pistidda, F. Sebastianutti, S. Lucca, G. Ramponi)

WHY?

The analysis of the system of property ownership in a city shows the balance between the public buildings and outdoor spaces and the private ones. The property regime also affects how buildings and outdoor spaces are used, their accessibility, their functions, and the different intervention methods. Knowing the public assets is fundamental to understanding how to plan the services of a city and how to correlate the properties with different regimes.

HOW

Building proper maps that contain all these categories is necessary to know the current state system of property ownership. The data can be collected from existing sources (existing official maps, official databases, official plans of the city, land and buildings registry, etc.) or verified/collected through surveys.

The thematic maps (Fig. 6.13) must include a legend showing the different categories of assets (for example, buildings and outdoor spaces) and the ownership regimes (public, semi-public, and private). The different types of properties will be identified through specific hatches.

Fig. 6.13 System of property ownership and accessibility of buildings and outdoor spaces. Conserving and reusing project of buildings and open spaces in Gyumri. Master "Training in Armenia in Restoration, 2011–2013", Politecnico di Milano, prof. M. Giambruno, R. Simonelli (© L. Khalatyan, S. Arakelian, R. Mkrtchyan)

RELATIONSHIP

The obtained data can be interlinked with other maps to realize cross-readings and thematic maps. For example, understanding the link between the different property regimes and the use of buildings and outdoor spaces can prove useful.

WARNINGS

These data are not easy to acquire. The documents in the city offices can be implemented with the information obtained from the interviews.

6.1.2.6 Building Materials, Paving of Outdoor Spaces, Elements of Urban Furniture

WHAT

Knowing the materials used for the buildings, the outdoor spaces, and the urban furniture is an important aspect of understanding the characteristics of a city's material heritage. The facades of the buildings with their materials and architectural elements, the different paving of the street and outdoor spaces and the different aspects of urban furniture create the urban structure and determine its quality. The survey of the building materials is also strictly correlated with their deterioration. It is necessary to outline a preliminary knowledge of the problems to be addressed while designing a conservation plan. This investigation and its output permit a thorough understanding of the buildings' conservation status.

This technical operation aims to identify the behavior of materials, study their alteration and degradation, and assess the conservation status.

WHY?

A materials survey is necessary to outline a preliminary knowledge of degradation to plan the intervention. It is one of the required steps to achieve the most accurate possible knowledge of the city and to understand the conservation issues to be addressed. It should focus on determining the causes of damage and decay. The material survey also defines the identity of urban space, and this analysis permits more conscious design choices.

HOW

The materials and construction techniques can be identified through attentive and repeated visual observations based on the experience gained in the field. The materials survey draws on the geometrical survey information to provide a complete understanding of the state of the elements of the urban spaces. A comprehensive visual analysis should lead first to the identification of the materials and then to their forms of alteration and decay. Identifying the materials using an appropriate classification system and nomenclature (ceramic materials; natural stones; artificial stones: mortar and concrete, brick and terracotta, glass, paintings; metallic materials; wood—it is important to identify the wood species; polymeric materials) is paramount. The data can be collected with the help of a data sheet subdividing the material according to the architectural elements (wall of the facades, windows and doors, decoration elements, roof, balconies, paving of outdoor spaces, etc.).

A detailed photographic campaign can partially integrate information not surveyed in situ. The materials survey is identified by mapping the drawings (Plan, 1:1000, 1:500) using selected hatches (Fig. 6.14).

RELATIONSHIP

The obtained data can be interlinked with other maps to realize cross-readings and thematic maps. For example, the materials survey must be linked with the decay survey to understand the causes and mechanisms of degradation.

WARNINGS

When the survey is conducted on an urban scale, it is advisable to organize the data collection method in advance with the help of cards and carefully decide the level of detail of the information to be collected.

Fig. 6.14 Building materials facades (left). The colors identify different types of masonry in the building facades (for example, the orange line represents stone masonry with traces of plaster, and the green line represents recent plaster). Outdoor spaces (right): paving of outdoor spaces, elements of urban furniture. The colored hatches identify the different paving materials: grey represents stone cubes, brown represents stone slabs, dark grey represents asphalt, green concrete, and yellow represents clay. The drawing also shows the presence of urban furniture. Fontecchio (L'Aquila, Italy), Reconstruction Plan, Politecnico di Milano (© M. Boriani, M. Giambruno, F. Maroldi, R. Simonelli, R. M. Rombolà, S. Pistidda, F. Sebastianutti, S. Lucca, G. Ramponi)

6.1.2.7 Characteristic Elements of Buildings and Urban Identity

WHAT

Knowing city history and historic buildings with their materials and characteristic elements allows us to understand the urban identity in a specific context.

Since the existing city is ever-changing, acknowledging this allows us to identify the elements of a collective identity. Knowing the features that contribute to creating an identity encourages one to respect the signs sedimented over time.

WHY?

An accurate survey of the main characteristic elements of buildings is useful to understand that the existing ones must be seen as archives because they enshrine different signs deposited by time and people. The survey is not just meant to take measures. Its aim is also to investigate, search for, know, and understand to gain awareness of the complex nature of architecture. Recognition and mapping of characteristic elements will be of great use in the control and authorization phases of restoration projects related to single buildings (Fig. 6.15).

HOW

The survey on the characteristic elements of buildings can be realized through attentive and repeated visual observations based on the experience gained in the field.

Fig. 6.15 Characteristic elements of buildings. Restoration and reuse project of the historical and architectural environment of the quarter in the central district of Yerevan, Armenia. Master "Training in Armenia in Restoration, 2011–2013", Politecnico di Milano, prof. M. Giambruno, R. Simonelli (© A. Avetisyan, L. Vardanian, N. Nalbandyan)

The survey draws on the geometrical survey information to provide a complete understanding of the localization of these elements. A comprehensive visual analysis should lead first to the identification of these elements and then to their state of conservation. The data can be collected with the help of a data sheet subdividing the elements according to their type (decoration elements, historical construction techniques, some specific use of historical material, etc.). A detailed photographic campaign can partially integrate information not surveyed in situ. The survey is identified by mapping the drawings (plans and elevations) using selected hatches or symbols. Maps on a scale of 1:1000 or 1:500 can show the characteristic elements with appropriate signs.

RELATIONSHIP

The obtained data can be interlinked with other maps to realize cross-readings and thematic maps. For example, this survey must be linked with the materials and decay survey to understand the state of conservation of the characteristic elements of the building.

WARNINGS

When the survey is conducted on an urban scale, it is advisable to organize the data collection method in advance with the help of cards and carefully decide the level of detail of the information to be collected.

6.1.2.8 Use of Public and Private Buildings and Outdoor Spaces (Ground Floor, Upper Floors)

WHAT

Public and private buildings and outdoor spaces in a city can have different types of use. Due to the different property regimes, it is also useful to identify the use categories of buildings and outdoor spaces to understand their functions, their accessibility, and whether there are any abandoned places needing renovation. Some examples of use categories of public spaces are services, green areas, sports centers, etc. Some examples of use categories of private spaces are residential use, commercial use, and tertiary use.

WHY?

The analysis of the use of public and private buildings and outdoor spaces in a city shows the balance between the different functions. A building can sometimes cover different functions: for example, the ground floor can often be used for different purposes from the upper floors (commercial use on the ground floor and residential use in the upper part of the building). The further use also affects the relationship between the building's ground floor, the street, and the upper floors. It is crucial to know the public assets to understand how to plan the services of a city and how to correlate the assets with different regimes. When designing in a historic context and re-functionalizing historic structures, the use of buildings and outdoor spaces must always be compatible with the historic pre-existence.

HOW

Building adequate maps with the identification of the different uses is fundamental to knowing the current use of the buildings and outdoor spaces, which is essential for the Master plan strategies.

The data can be collected from existing sources (existing official maps, official databases, official plans of the city, land and buildings registry, etc.) or verified/ collected through surveys. The thematic maps must include a legend with the different categories of assets (for example, public and private buildings and outdoor spaces) and further use (residential, commercial, gardens, etc.). The different types of use will be identified through specific hatches.

RELATIONSHIP

The obtained data can be interlinked with other maps to realize cross-readings and thematic maps. For example, can be useful to understand the link between the use of the buildings and outdoor spaces and the different property regimes. It is also important to study the various functions on the ground floor, because it establishes a specific relationship between the built and outdoor spaces, influencing architectural choices (dimension of the street, presence of fences, use of greenery in the outdoor spaces, etc.).

WARNINGS

A map of the ground floors can be useful to connect the uses of the indoor and the outdoor spaces.

6.1.2.9 Improper or Spontaneous Use, Buildings in a State of Ruin or Abandonment, Additional Storeys, Informal Buildings, and "Domik"

WHAT

In some contexts, the urban texture is very complex and may present buildings and outdoor spaces used improperly or spontaneously, buildings in a state of ruin or abandonment, additional storeys and informal buildings, and in the case of Gyumri, also "domik". All these elements and stratifications characterize the urban context, and it is necessary to evaluate which ones can turn into opportunities and which ones are critical issues. The knowledge and an appropriate survey of these elements permit to understand the place's identity in all its specificities.

WHY?

The improper or spontaneous use of buildings or outdoor spaces, buildings in a state of ruin or abandonment, additional storeys, informal buildings and "domik" are part of the historical stratification of the city. They are often the expression of informal community interventions. It is very important to know all these elements and their relationship with the built environment to understand how to reuse them or whether they are critical points that can be demolished. Buildings in a state of ruin or abandonment can become an opportunity in the city's strategic plan. They can be reused with new functions, considering that it is important to foresee a compatible use with the existing structures.

HOW

In order to know where buildings or outdoor spaces are used improperly or spontaneously and where buildings in a state of ruin or abandonment, additional storeys, informal buildings, and "domik" are, it is necessary to create adequate maps in which these elements may be localized. The data can be collected from existing sources (existing official maps, official databases, official plans of the city, land and buildings registry, etc.) or verified/collected through site surveys. The thematic maps must include a legend with the different categories of informal use that can be identified through specific hatches. A detailed photographic campaign can partially integrate information not surveyed in situ.

RELATIONSHIP

The obtained data can be interlinked with other maps to realize cross-readings and thematic maps.

WARNINGS

Data about improper uses are sensitive data and can sometimes be obtained only by involving the inhabitants. The survey of these data requires particular attention.

6.1.2.10 Services, Trade, and Activities

WHAT

After mapping the uses of public and private buildings and outdoor spaces, it can be interesting to add details with more information about the specific use of these places. For example, commercial use can be documented by collecting detailed information about which activity characterized a street or some building or outdoor spaces (grocery stores, clothing stores, personal service, etc.). Each macro function can be detailed by identifying a specific vocation of use in the city and in the neighborhoods.

WHY?

The analysis of services, trade, and activities can identify a part of the city characterized by a specific vocation of use or underline some missing functions. This kind of mapping is very important to understand the strategy of a new plan and decide how to concentrate or distribute the individual functions in the city. Understanding the right balance of the activities to reuse the urban context and maintain its identity becomes of crucial importance.

HOW

To know the current type of services, trade, and activities of buildings and outdoor spaces it is necessary to create adequate maps in which the different categories may be recognized. The data can be collected from existing sources (existing official maps, official databases, official plans of the city, land and buildings registry, etc.) or can be verified/collected through surveys.

The thematic maps must include a legend with the different categories of assets (residential use, commercial use, services) and the details of other activities (residential for students, grocery stores, clothing stores, personal services, hospitals, schools, etc.). The different types of services, trade, and activities will be identified through specific hatches.

RELATIONSHIP

The obtained data can be interlinked with other maps such as property, uses of the outdoor spaces and buildings.

WARNINGS

A detailed survey can require a lot of time, so it is necessary to identify the aims of this reading in advance.

6.1.2.11 Expeditious Conservation Status of Buildings and Public and Private Outdoor Spaces

WHAT

The knowledge of the conservation status of buildings and public and private outdoor spaces is an important aspect of understanding the conditions of the built heritage and the outdoor spaces. When accessible, it is possible to integrate the information with data about indoor conditions.

WHY?

A first analysis of the level of decay of buildings and outdoor spaces is necessary to outline a preliminary knowledge of the decay problems on an urban scale, which is paramount to planning the interventions (evaluation of the conservation status, the ongoing decay phenomena, and their causes).

HOW

The survey of the synthetic conservation status could be done with proper data sheets drawn up in relation to the specific situation. The data concerning a first synthetic conservation status can be processed in maps and data sheets that explain the different levels of decay of the built environment and outdoor spaces.

To facilitate an overall reading and provide useful interpretations to draft the working Plan, decay phenomena levels can be grouped into different categories related to the gravity of the damage caused to the building (for example, no worse conditions, formation of secondary products, and loss of materials). For the outdoor spaces, the paving status, the presence of urban furniture, the presence of lighting, and the quality of greenery are some of the elements to be analyzed (Fig. 6.16).

RELATIONSHIP

The synthetic conservation status of the building and public spaces can be interlinked with other maps to realize cross-readings. For example, the conservation status must be linked with the materials survey to understand the causes and mechanisms of degradation and with the maps about the use of the building to understand whether it is in use.

Fig. 6.16 Conservation status of buildings and outdoor spaces. Restoration and reuse project of the historical and architectural environment of the quarter in the central district of Yerevan, Armenia. Master "Training in Armenia in Restoration, 2011–2013", Politecnico di Milano, prof. M. Giambruno, R. Simonelli (© A. Avetisyan, L. Vardanian, N. Nalbandyan)

WARNINGS

The levels of decay are not standard, but they can be identified according to the general conditions of the context.

6.1.2.12 Conservation Status of the Building Facades

WHAT

After the first synthetic conservation status of buildings and public and private outdoor spaces, it is important to detail the knowledge on the conservation status of the building facades. The survey on the decay of the building facades is necessary to outline a more detailed understanding of the problems to be addressed while designing the conservation project. This investigation and its output are drawn over a geometrical survey, which permits a thorough knowledge of the buildings' conservation status. The data can be processed in some drawings with a mapping of the prospects that identify the most important decay phenomena. Each decay phenomenon can also be explained in specific data sheets relying on images that define issues, causes, and the relationship with the materials.

WHY?

A decay survey is necessary to outline a preliminary knowledge of degradation to propose the intervention category in the Urban Plan. It should focus on determining the causes of damage and decay.

HOW

The identification of decay phenomena must be realized through attentive and repeated visual observations based on the experience gained in the field, using a glossary as an international support and referring to the ICOMOS-ISCS: Illustrated glossary on stone deterioration patterns. A detailed photographic campaign can partially integrate information not surveyed in situ. The deterioration phenomena are identified by mapping the deteriorated surfaces that can be represented on the drawings (prospects and sections) using selected hatches.

Usually, the decays inspected in situ should be represented on the geometrical survey or orthophoto in scale 1:500, 1:200 (Fig. 6.17) for an urban scale (metric decimal system)—except in cases where a significant level of detail is requested—by using symbols specifically studied to be readable and interpretable at that scale.

The realization of datasheets presenting the description of each deterioration phenomenon identified and the individuation of its possible causes can be useful to improve the results and favor a better understanding of the deterioration survey.

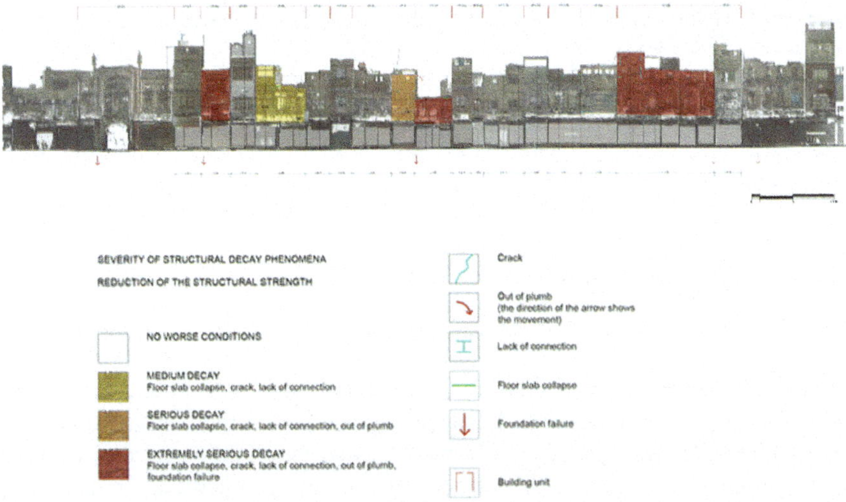

Fig. 6.17 Conservation status of the building facades. Conserving and reusing project of the buildings and outdoor spaces in Multan, Pakistan. Sarafa Bazaar regeneration, Sustainable Social Economic and Environmental revitalization in the historic core of Multan city, Pakistan (© M. Boriani, M. Giambruno, C. Manfredi, R. Mastropirro, R. Gabaglio, S. Pistidda, M. Tasinato)

RELATIONSHIP

Materials and decay phenomena survey should be realized together with the study of the structural problems and possibly with further chemical, physical, structural and geological analyses that may be needed case by case.

Structural and surface decays could only be correctly interpreted in an overall reading context since they are not exclusively dependent on the conditions and peculiarities of every single building but are subject to environmental conditions.

WARNINGS

When the survey is conducted on an urban scale, it is advisable to organize the data collection method in advance with the help of data sheets and carefully decide the level of detail of the information to be collected.

6.1.2.13 Datasheet for a Rapid Survey of Buildings and Outdoor Spaces (Sheet Outline)

WHAT

The datasheets for the rapid survey are an essential tool to develop the knowledge process on the buildings and outdoor spaces. Primarily, when the work must be implemented at an urban level by collecting a significant amount of information, a practical tool is necessary to facilitate the process of acquisition and restitution in a short time. The tool can be organized in a simple way to be filled quickly, even by unskilled staff.

WHY?

Any plan for historic cities must be preceded by careful studies about the consistency of the built heritage and the outdoor spaces. This information is fundamental to understanding the present conditions at different levels to orient the Plan's strategies.

HOW

The datasheets (Fig. 6.18) allow the organization of the information in thematic areas, and a multiple-choice tool guarantees quick and easy compilation. It is possible to collect data about the outdoor spaces according to their morphological nature (presence of streets, squares, alleys), their use (vehicular, pedestrian), their materials (type of paving), and their equipment (parking, street furniture, green spaces, lighting system). The data sheets on the buildings can give information about the exterior and the interior parts in case of accessibility. Valuable data can be related to the property (private or public), the uses (residential, commercial, public, etc.), the characters of the buildings (number of floors, related outdoor spaces, added elements), the materials, and the conservation status.

Fig. 6.18 The datasheets developed for the first survey in Gyumri for the ADB consulting, March 2022 (elaboration by authors)

RELATIONSHIP

The information collected through the datasheets can be managed with a G.I.S. system to extract thematic readings. Such a tool can be a useful system to manage the Plan before, during and after its implementation.

WARNINGS

The datasheets are not a standard tool, but they must be implemented according to the peculiarities of the site. Therefore, the tool requires a careful preliminary design after some site survey and initial knowledge of the context.

6.1.2.14 Datasheet for the Survey of Buildings Interiors Conditions (Sheet Outline)

WHAT

he datasheets for the survey of the buildings' interior conditions (Fig. 6.19) allow for the completion of the knowledge process at every level of study. The connection with the data collected on the exteriors at an urban level provides a complete overview of the conditions of all urban blocks (Fig. 6.20).

WHY?

The information related to the buildings' interiors is important to simultaneously manage the urban level and the building level.

HOW

The census of the interiors provides an overview of the buildings' characters and conditions and defines priorities for the Plan. The sheets can include information about the interior distribution, the number of people, ownership, equipment, materials, and characters, the presence of historical elements or alteration works of the original features, state of preservation, and living conditions.

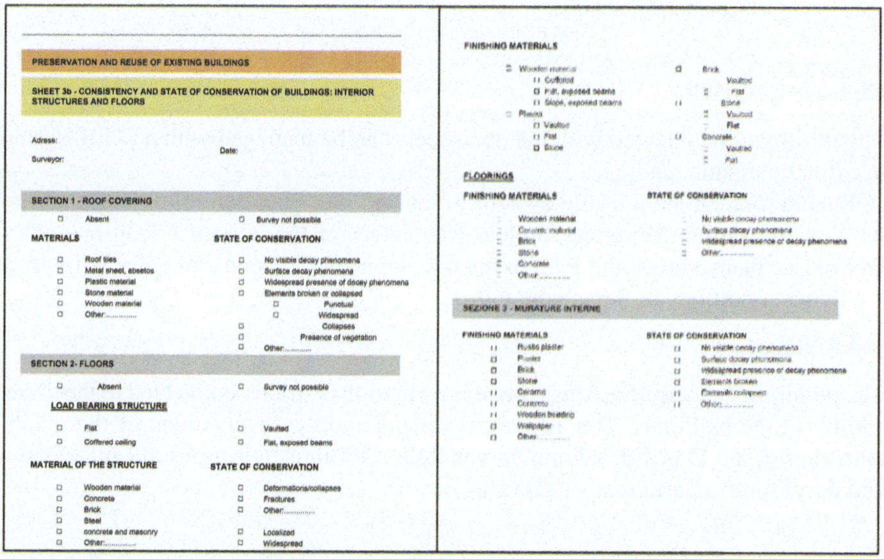

Fig. 6.19 An example of the datasheet for the survey of the buildings' interiors, structures and floors (elaboration by authors)

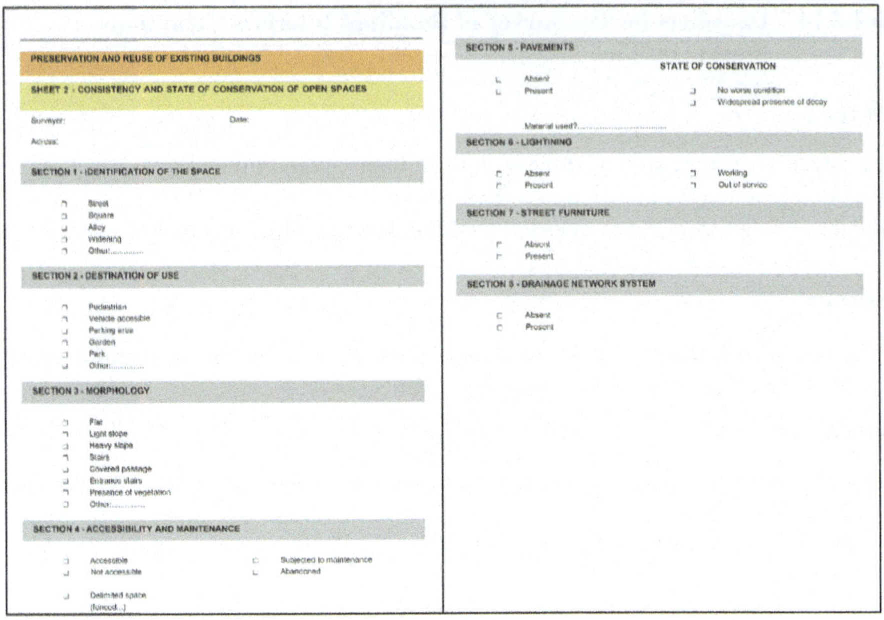

Fig. 6.20 An example of the datasheet for the survey of consistency and state of conservation of open spaces (elaboration by authors)

RELATIONSHIP

The information collected with the datasheets can be managed with a G.I.S. system to extract thematic readings.

Such a tool can be a useful system to manage the Plan before, during and after its implementation. Moreover, a clear framework of the interior conditions of the buildings can also guide the Plan in the development of social strategies to improve the living conditions of the inhabitants.

WARNINGS

The possibility of acquiring this type of information is strictly connected to the accessibility of the buildings. The awareness-raising and the involvement of the inhabitants during the Plan's development can help a lot in acquiring useful information and developing a participatory process.

6.1.2.15 Questionnaires: Inhabitants' Perception, Needs, and Requirements of the Historic Center and Its Critical Issues (Questionnaires Outline)

WHAT

Questionnaires are simple tools to mobilize the inhabitants to contribute to the discussion in the implementation of the Plan. It's also a tool for comparing, verifying, and deepening the information collected with the datasheets to enhance the level of knowledge (Fig. 6.21).

WHY?

Listening to the needs of the inhabitants is an essential step in the entire knowledge process. The collected data can give useful information to better orient the Plan's strategies and identify priorities or problems not visible from the site survey alone. The voice of the site's primary users can give essential data to identify problems and perspectives for urban regeneration.

HOW

The section of the questionnaires can be organized with different data: some questions can be addressed to understand the profile of the inhabitants of the neighborhoods (age, level of education, employment), the provenience, the level of perception of the importance of the places, their expectations and their will to keep living there or not. The questionnaires are structured with YES and NO modalities or require a short description, and they can be completed through direct interviews with the inhabitants or through social media.

RELATIONSHIP

The questionnaires can be targeted to specific users to collect different types of information: inhabitants, workers, administrators, professionals, etc.

WARNINGS

Engaging the inhabitants is not an easy task. The interviewers must consider that different people's targets require different ways of involvement. The questionnaire must be designed in a simple way to obtain as many answers as possible—and thus representative results.

6.1.2.16 Analysis Summary: Identification of Criticalities, Priority Areas of Intervention, and Areas Needing Primary Protection

WHAT

The analysis summary synthesizes and schematizes all the results of the knowledge process. It represents the preliminary base for identifying criticalities, problematic

QUESTIONNAIRE - Kumayri historic district – Gyumri

Section 1: General profiling information:

1) What is your age range?

- <15 years old
- 16-20 years old
- 21-25 years old
- 26-34 years old
- 35-44 years old
- 44-60 years old
- >60 years old

2) What is your level of education?

- Primary school certification
- Secondary / High school diploma
- University degree
- Post-university qualification (Post graduate Master, PhD).

3) What is your employment status?

- Employed
- Unemployed
- Student
- Retired
- Other (please, specify: ________________)

Section 2: Gyumri and Kumayri historic district

4) Do you live in Gyumri?

- Yes
- No

4a) If NO to question 4, where do you come from?

- City of the same province
- City of another province
- Village of the same province
- Village of another province

4b) If YES to question 4, do you want to continue to live here?

- Yes
- No

4c) If NO to question 4, why?

- Lack of services
- Lack of work possibilities
- Lack of housing in good conditions

4d) If YES to question 4, do you live in Kumayri historic district?

- Yes
- No

4e) If NO to question 4, for which reasons do you visit Kumayri historic district? (multiple choices allowed)

- Work
- Study
- Free time (cultural visits, shopping...)
- Vacation

Section 3: Characters of Kumayri historic district – Gyumri (section open to inhabitants, workers, students, visitors)

5) Do you think that Kumayri historic district is different from other districts of Gyumri?

- No
- Not so much different
- Quite different
- Yes

6) In your opinion, what makes Kumayri historic district different from other districts of Gyumri? (multiple choices allowed)

- The presence of historic buildings
- The presence of cultural spaces (museums, libraries...)
- The presence and frequency of monuments
- Other (please, specify: ________________)

7) In your opinion, which are the aspects of the Kumayri that need to be improved? (multiple choices allowed)

- An active legislation: regulation specifically indicating old buildings and new buildings;
- Definition of the characters that define the historical importance of the buildings and so to be preserved; indicate precise rules for example for the colors of the facades, for the materials to be used etc. etc. (e.g. identify historic buildings and create a pilot project for those buildings)
- A list of what the private can do, in order to create a close collaboration between public and private
- Care must be managed by the public and not left to the goodwill of the private sector: close collaboration between public and private to make care effective
- The condition of historic buildings
- The cultural spaces (museums, libraries...)
- The aggregative spaces
- The presence and quality of public spaces (parks, gardens)
- The quality and condition of streets
- The frequency and quality of commercial activities
- Active presence of the institutions
- Availability of public transportation means
- Availability of public services (e.g. pharmacies, post offices, ATM...)
- Availability of social housing
- Removal and storage of toxic materials (e.g. eternit on the roofs)
- Garbage cycle (e.g. separate garbage collection, etc.)
- The public lighting system
- Working opportunities
- Tourism facilities
- Other (please, specify: ________________)

8) In your opinion, which parts of Kumayri historic district needs more improvement?

8a) In your opinion, is it necessary to intervene on Vartanants Square?

- Yes
- No

8b) If YES to question 8a, which interventions are needed in your opinion?

- Lighting system
- Furniture (benches, public toilets)
- Trees – green areas
- Regulating parking
- Creation of pedestrian areas
- Small shops
- Improvement of the facades
- Other (please, specify________________________)

Section 4: Perception of living in Kumayri historic buildings – (section open only to inhabitants)

9) Do you live in a historic building? (end of XIX century – 1960s)

- Yes
- No

10) If YES to question 9, what are the conditions of your building?

- Good conditions
- Fair conditions
- Bad conditions

10a) If YES to question 10, have you ever made maintenance works on your building?

- Yes
- No

10b) If you answered to question 10a, did you make the intervention by yourself?

- Yes
- No

10c) If YES to question 10b, which maintenance works have you done?

- Repaired the roof
- Repaired cracks on masonry
- Added floors
- Rebuilt collapsed parts
- Changed interiors layout
- Added sanitation systems
- Changed windows and door frames
- Changed the electrical / sanitation plant
- Other (please, specify___________________)

11) Do you think that guidelines for the self maintenance and conservation of your building would be useful?

- Yes
- No

12) Which possible development do you imagine for Gyumri? (open question)

Fig. 6.21 The questionnaires developed for the first survey in Gyumri for the ADB consulting, March 2022. The tool was discussed in Gyumri and revised with Doctor Antonio Montalto, whom we thank for his precious contribution (elaboration by authors)

Fig. 6.22 Map of the resources and criticalities of the territory, Fontecchio (L'Aquila, Italy), Reconstruction Plan, Politecnico di Milano (© M. Boriani, M. Giambruno, F. Maroldi, R. Simonelli, R. M. Rombolà, S. Pistidda, F. Sebastianutti, S. Lucca, G. Ramponi)

areas, resources and elements to be protected. These data constitute the basic tool to set priority areas of intervention and areas needing maximum protection (Fig. 6.22).

WHY?

The knowledge process needs to be synthesized critically, to be useful and effective. The summary can give the preliminary indications to start the development of the Plan.

HOW

The analysis summary can follow different deepening levels based on the cartography elaborated in the preliminary phase. They start from the territorial scale (1:2000, 1:1000) to enter a more detailed scale (1:500) for specific needs. The criticalities, priorities of intervention and other plans' reflections can be represented in the same map with hatches and patterns to give a general overview of all the issues. The use of the G.I.S. system in all the work steps can help manage all the data.

RELATIONSHIP

All the knowledge process finds its synthesis in this final document, so it is suggested to share the results with all the people involved in the first step: inhabitants, professionals, administrators, etc.

WARNINGS

This document is the final step of the knowledge process and the beginning of the Plan proposals. It requires a careful moment of discussion between all the stakeholders involved to share their reflections.

References

1. Caccia Gherardini S (2016) The architectural restoration and conservation handbook. Didapress, Firenze
2. Feilden BM, Jokilehto J (1998) Management guide for cultural heritage site. ICCROM, Roma

SPRINGER NATURE

GPSR Compliance

The European Union's (EU) General Product Safety Regulation (GPSR) is a set of rules that requires consumer products to be safe and our obligations to ensure this.

If you have any concerns about our products, you can contact us on ProductSafety@springernature.com

In case Publisher is established outside the EU, the EU authorized representative is:

Springer Nature Customer Service Center GmbH
Europaplatz 3
69115 Heidelberg, Germany

The manufacturer's authorised representative in the EU is Springer Nature Customer Service Centre GmbH, Europaplatz 3, 69115 Heidelberg, Germany. If you have any concerns regarding our products, please contact ProductSafety@springernature.com

Printed and bound by CPI Group (UK) Ltd, Croydon, CR0 4YY
30/09/2026
02225474-0006